A
FIRST COURSE IN
STATISTICS

A
FIRST COURSE IN
STATISTICS

BY

ROBERT LOVEDAY

M.Sc., F.I.S.

Principal Lecturer in Mathematics and Statistics,
The Technical College,
Kingston-upon-Thames

CAMBRIDGE
AT THE UNIVERSITY PRESS
1964

CAMBRIDGE
UNIVERSITY PRESS

University Printing House, Cambridge CB2 8BS, United Kingdom

Cambridge University Press is part of the University of Cambridge.

It furthers the University's mission by disseminating knowledge in the pursuit of education, learning and research at the highest international levels of excellence.

www.cambridge.org
Information on this title: www.cambridge.org/9781316607008

First edition 1958
Reprinted 1960, 1961, 1964
First paperback edition 2016

A catalogue record for this publication is available from the British Library

ISBN 978-1-316-60700-8 Paperback

Contents

CONTENTS

vi

CONTENTS

CONTENTS

Chapter 7. CORRELATION BY RANKS

Chapter 8. THE ANALYSIS OF A TIME-SERIES

Chapter 9. WEIGHTED AVERAGES

Chapter 10. MISCELLANEOUS TOPICS

CONTENTS

Contents

Preface

The title chosen for this book might have been 'Stage A Statistics', since, like Stage A Geometry, it is numerical, experimental and practical. It is designed not only to meet the requirements of students preparing for Statistics as an Ordinary Level subject of the G.C.E., but also to give a useful and interesting course of study to boys and girls in Technical Schools. It may also prove helpful to students in Teachers' Training Colleges and to students of Medicine, Dentistry, Agriculture, Economics or Engineering who require an elementary introduction to the subject before embarking upon a more mathematical treatment. It contains many examples and exercises, mainly taken from the past examination papers of the Cambridge Local Examinations Syndicate, the London University G.C.E. and the Northern Universities' Joint Matriculation Board. The source of each is shown, and the author's thanks are due for permission to reproduce them. Use has also been made of data taken from *The Annual Abstract of Statistics* and *The Monthly Digest of Statistics* published by H.M. Stationary Office.

Five more difficult sections and exercises, marked with an asterisk, may be omitted on a first reading (42, 43, 57, 61, 62).

'A Second Course in Statistics', will follow. This will be an Advanced and Scholarship Level treatment of the subject. It will introduce probability, the binomial distribution, the normal distribution, significance and confidence limits. It will deal also with regression by the 'least squares line of best fit' and give a full treatment of the product-moment correlation coefficient and its significance.

A First Course in Statistics is dedicated to all those boys and girls who have worked under my guidance during the past twenty years. Their difficulties have shown me which points it is necessary to labour, and how far to take the subject.

R. L.

1 January 1958

1

Frequency Distributions

1. The formation of a frequency distribution. Suppose there are thirty-six tomato plants in a garden each bearing three trusses of fruit and that we go round the plants counting the number of tomatoes per truss. The results of our survey might be recorded as in table 1 A.

TABLE IA

No. of tomatoes per truss					Frequency
8	1				1
9	1				1
10	11				2
11	ﬀﬀﬀ				5
12	ﬀﬀﬀ	1111			9
13	ﬀﬀﬀ	ﬀﬀﬀ	ﬀﬀﬀ		15
14	ﬀﬀﬀ	ﬀﬀﬀ	ﬀﬀﬀ	1111	19
15	ﬀﬀﬀ	ﬀﬀﬀ	ﬀﬀﬀ	ﬀﬀﬀ	20
16	ﬀﬀﬀ	ﬀﬀﬀ	ﬀﬀﬀ	1	16
17	ﬀﬀﬀ	ﬀﬀﬀ			10
18	ﬀﬀﬀ				5
19	111				3
20	1				1
21					0
22	1				1

Total number of trusses examined   108

It should be noted that a small stroke is placed in the appropriate row of the middle column as each truss is counted and that every fifth stroke is drawn diagonally across the preceding four strokes to form small groups of five. This enables us to see at a glance the total for each row when we enter it in the right-hand column giving the *frequency* with which each particular number of tomatoes per truss occurs. Table 1 A is an example of a *frequency distribution*.

2. The histogram. It is customary to represent a frequency distribution diagrammatically in the form of a *histogram* as shown in fig. 1.

3. The mode. The size of truss which occurs most frequently is 15, and this is called the *mode*.

4. A frequency distribution by grouping. Let us next suppose that, while gathering in the potato crop, we weigh in lb. and oz. the quantity of potatoes obtained from each of 100 roots and that our results are shown in table 1 B.

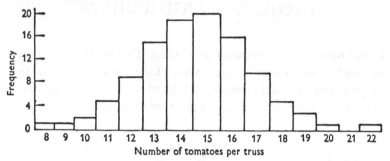

Fig. 1. Histogram.

TABLE 1B

Weight of potatoes per root in lb.		Frequency
Under 3	‖‖ 1111	9
3–	‖‖ ‖‖ ‖‖ ‖‖ 11	22
6–	‖‖ ‖‖ ‖‖ ‖‖ ‖‖ 111	28
9–	‖‖ ‖‖ ‖‖ ‖‖ 1	21
12–	‖‖ ‖‖ ‖‖ 11	17
15–18	111	3
	Total	100

In this case the weights in the left-hand column are grouped into classes of 3 lb. intervals the first row showing the number of roots yielding under 3 lb. of potatoes, the second row showing the number of roots yielding from 3 to 6 lb. but not including 6 lb. and so on. The histogram for this frequency distribution would be as shown in fig. 2.

5. Modal class. The 6–9 lb. class in the distribution shown in fig. 2 is called the *modal class*. In the first example the size of truss occurring most frequently was called the mode. In this example, however, it is quite probable that no two of the 100 roots have exactly the same weight, and there may be no actual weight which occurs most frequently. The name *modal class* is therefore given to the *class* which contains the greatest number of members, that is, to the 6–9 lb. class.

6. Discrete and continuous variation. It should be noted that the number of tomatoes per truss varies *discretely*, whilst the weight of potatoes per root varies *continuously*. The weight of a root of potatoes might have any value between 0 and 18 lb. It need not be a whole number. It might lie, for example, somewhere between 6 and 7 lb., while the number of tomatoes must be 6 or 7.

7. The frequency polygon. In fig. 2 the polygon *ABCDEF*, whose sides are the straight lines joining the mid-points of the tops of the rectangles of the histogram, is called the *frequency polygon*. Several frequency polygons can be drawn on one diagram, thus enabling us to compare frequency distributions. An illustration of this is given in §8.

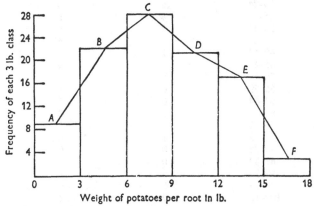

Fig. 2. Histogram and frequency polygon.

8. Example. An analysis of the number of words per sentence in the first hundred sentences of

    (*a*) *Pride and Prejudice*, by Jane Austen,

    (*b*) *The Cathedral*, by Hugh Walpole,

gives the frequency table as shown on p. 4.

On one sheet of graph paper draw two frequency polygons, setting them out in such a way that you can use them to compare the two distributions given in the table.

Discuss the use of this analysis to demonstrate that one of the novels is more difficult to read than the other. (Northern.)

The two frequency polygons shown in fig. 3 indicate quite clearly that Hugh Walpole's sentences are longer than Jane Austen's, the modal group of *The Cathedral* being 15–19 words per sentence, that of *Pride and Prejudice* 5–19 words per sentence. This might demonstrate that *The Cathedral* is more difficult to read than *Pride and Prejudice* were it not for the following important factors which are not shown in the graphs:

    (i) Difficulty of vocabulary.

    (ii) Does the author get the reader's interest?

No. of words per sentence	No. of sentences	
	(a) *Pride and Prejudice*	(b) *The Cathedral*
0–4	6	2
5–9	33	12
10–14	22	14
15–19	15	19
20–24	10	18
25–29	4	6
30–34	2	5
35–39	3	6
40–44	2	6
45–49	2	4
50–54	0	1
55–59	0	1
60–64	0	2
65–69	0	0
70–74	0	1
75–79	1	0
80–84	0	0
85–89	0	0
90–94	0	1
95–99	0	0
100–104	0	2

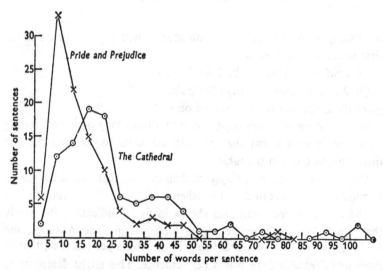

Fig. 3. Two frequency polygons in one diagram.

4

(iii) The first hundred sentences form a very small sample of the whole book.

(iv) They may not be typical of the whole book.

This example shows that the greatest care must be exercised in the drawing of conclusions from statistical data.

9. Experiments with dice. The histograms or frequency polygons obtained from the measurements of natural phenomena generally approximate in shape to certain well-defined curves called *frequency curves*. It is an amusing and useful exercise for the beginner to establish the shapes of these curves by throwing dice.

10. Throwing three dice. Normal distributions. It can be shown mathematically that, if three dice are cast, the probability of getting three sixes is 1/216. This means that there is 1 chance in 216 of scoring a total of 18. Further, the probability of getting two sixes and one five is 3/216. This means that there are 3 chances in 216 of scoring a total of 17. The following table is a complete list of the chances for the various possible total scores:

Total score	3	4	5	6	7	8	9	10
Chances in 216 of getting the score	1	3	6	10	15	21	25	27
Total score	11	12	13	14	15	16	17	18
Chances in 216 of getting the score	27	25	21	15	10	6	3	1

Fig. 4a shows the frequency curve obtained by drawing a graph from the above table. If the student actually throws three dice a large number of times and records the scores as shown in table 1 A he will obtain a frequency distribution which will give a histogram or frequency polygon approximating in shape to the *normal frequency curve*. The author supervised the throwing of three dice 4320 (i.e. 20×216) times by twenty boys and obtained the following distribution, the frequency polygon of which approximates very closely in shape to fig. 4a:

Score	3	4	5	6	7	8	9	10
Frequency	19	52	93	194	285	411	473	510
Score	11	12	13	14	15	16	17	18
Frequency	541	520	438	333	218	149	68	16

11. Throwing six dice. Skewed distributions. If six dice are cast, it can be shown by the binomial theorem that the probability of getting no sixes is approximately 72/216, while that of getting one six is approximately 87/216. These and other values have been used in the drawing of fig. 4*b* which shows the shape of curve associated with a *skewed* distribution. When the greater part of the curve exists to the right of the mode it is said to be skewed positively. The student will also meet cases of

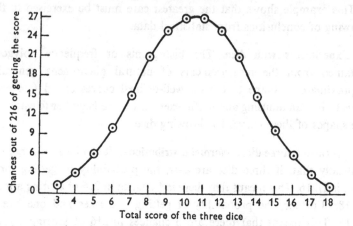

Fig. 4*a*. Approximately normal distribution. (A truly normal curve is rather more sharply peaked rising to 29 instead of 27. It will be fully discussed in 'A Second Course in Statistics'.)

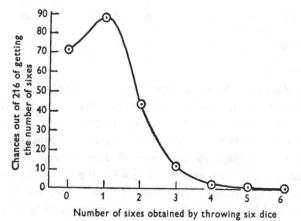

Fig. 4*b*. Skewed distribution (skewed positively).

negative skewness in which the greater part of the curve is to the left of the mode. *Skewness* is, therefore, a term which distinguishes distributions from *normal* distributions. The latter are symmetrical about the mode.

The following distribution was obtained in an actual experiment:

No. of sixes	0	1	2	3	4	5	6
Frequency	77	84	34	19	2	0	0

6

The student will find the frequency polygon obtained from this distribution closely resembles fig. 4*b*, but he should carry out a similar experiment himself.

12. Throwing one die. Rectangular distributions. If a single die is cast, each number has an equal chance of appearing uppermost. Thus the probability for each number is 1/6. This can be stated quite simply as 1 chance in 6, but in fig. 4*c* it has been interpreted as 36 chances in 216 in order to keep the totals the same in figs. 4*a*, *b* and *c*. Thus the

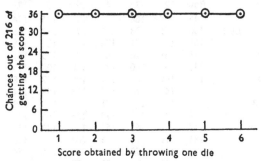

Fig. 4*c*. Rectangular distribution.

frequency curve in this case is a *straight line parallel to the abscissa of the graph* and the distribution is described as *rectangular*. The following table gives the results of an actual experiment:

Score	1	2	3	4	5	6
Frequency	94	96	115	140	105	98

The student should carry out a similar experiment for himself. In these experiments it is assumed, of course, that the dice are true (i.e. that they are not loaded).

13. Exercises.

1. The following are the populations, in hundreds, of 80 towns and villages in England and Wales, taken at random from a hotel directory (towns of over 50,000 inhabitants are omitted):

11	200	31	7	11	15	17	55
72	6	72	13	52	377	12	9
15	16	18	5	139	18	30	3
8	6	30	142	159	46	171	149
15	131	43	70	89	1	30	25
3	5	2	86	4	24	6	52
23	18	1	31	52	48	160	12
26	407	52	38	28	5	58	124
2	99	40	70	19	26	222	120
319	127	3	51	150	39	13	10

7

Form a frequency distribution and construct the histogram. The following grouping is appropriate: 0–29, 30–59, 60–89, etc. State which is the modal group. (Northern.)

Note that the mode is 52, since it occurs in the list 4 times and no other number occurs more than 3 times.

2. *Cambridge University Boat Club*

Weights of crews, in lb. (*to nearest lb.*), *in the Lent Bumping Races,* 1956

First Division

	Bow	2	3	4	5	6	7	Str.	Cox
Jesus 1	148	168	168	174	172	185	174	174	132
1st and 3rd Trinity 1	160	171	178	157	178	183	168	168	133
L.M.B.C. 1	161	162	171	173	156	181	168	155	128
Peterhouse 1	171	168	176	169	196	175	157	171	125
Pembroke 1	146	152	160	170	190	190	169	173	133
Emmanuel 1	145	169	165	174	199	182	178	170	122
Clare 1	154	144	178	194	168	187	144	154	124
Trinity Hall 1	163	165	168	171	184	172	169	184	136
King's 1	161	146	170	173	166	161	172	160	126
Corpus Christi 1	147	160	161	153	174	187	152	150	157
Magdalene 1	183	179	174	175	209	190	196	175	140
Jesus 2	173	164	178	169	209	164	154	145	134
Queens' 1	150	171	176	182	178	176	175	171	116
Christ's 1	148	168	176	178	183	186	170	159	138
Caius 1	161	165	173	182	186	176	177	202	118
St Catharine's 1	143	169	174	171	181	167	168	146	148

Second Division

	Bow	2	3	4	5	6	7	Str.	Cox
1st and 3rd Trinity 2	146	147	164	168	170	158	154	154	126
St Catharine's 2	156	164	178	148	182	158	186	154	136
Selwyn 1	164	160	159	193	188	158	164	185	127
L.M.B.C. 2	158	163	164	179	178	172	177	163	128
Downing 1	141	157	154	185	191	178	168	166	124
Sidney Sussex 1	152	154	189	176	177	189	187	189	130
Clare 2	178	147	170	151	189	170	182	156	112
Trinity Hall 2	161	150	160	178	173	175	161	162	124
Jesus 3	158	163	159	168	164	178	172	161	132
Fitzwilliam 1	156	181	203	208	169	190	151	155	123
Pembroke 2	162	156	177	166	195	175	153	147	136
Peterhouse 2	150	140	157	184	220	169	140	171	126
1st and 3rd Trinity 3	154	160	162	174	159	178	161	146	128
Emmanuel 2	150	136	149	185	150	224	164	152	134
Caius 2	154	158	161	178	171	166	174	147	125
King's 2	152	156	173	167	159	163	170	136	134

Form a frequency distribution for the above table which shows the weights of 288 men and construct the histogram. The following grouping is appropriate: $109\frac{1}{2}$–$119\frac{1}{2}$, $119\frac{1}{2}$–$129\frac{1}{2}$, $129\frac{1}{2}$–$139\frac{1}{2}$, etc., to $219\frac{1}{2}$–$229\frac{1}{2}$. The cox's constitute a different *weight population* to the rest of the crews, and it will be noticed that, if the weights of the coxes are excluded the frequency polygon is *approximately* normal in shape. The inclusion of the coxes causes it to be *skewed negatively.*

Which is the modal class?

3. The following is a record of marks obtained by a group of boys in an examination:

25	63	82	71	12	57	63	38	17	23	0	96	81	72	35	76	44
54	19	70	45	70	44	43	18	93	2	15	60	71	82	3	61	64
25	42	40	70	63	62	83	18	27	58	50	52	53	89	90	50	53
23	81	70	58	31	32	28	19	23	72	58	37	33	30	20	62	71
48	63	62	59	38	35	37	46	81	73	75	63	65	53	47	52	38

Tabulate as a frequency distribution, grouping in intervals of 10 marks 0–9, 10–19, 20–29, etc. Plot the result as a histogram and determine the modal class.

By examining the original marks determine the mode. (London.)

4. Make out a table similar to the following and use it to take a census of the sizes of caps worn by the boys of your school:

	Size of cap										
Form	$6\frac{1}{4}$	$6\frac{3}{8}$	$6\frac{1}{2}$	$6\frac{5}{8}$	$6\frac{3}{4}$	$6\frac{7}{8}$	7	$7\frac{1}{8}$	$7\frac{1}{4}$	$7\frac{3}{8}$	$7\frac{1}{2}$
I*a*	1	4	8	7	6	3	1	.	.	.	.
I*b*	.	3	5	7	8	6	0	1	.	.	.
II*a*	.	2	6	9	8	3	1	1	.	.	.
etc.	.	.	.	.	.	.	.	.	.	.	.
Totals for each size for the whole school	1	9	19	23	22	12	2	2	.	.	.

Draw the histogram and state the modal size of cap.
Explain briefly how the school outfitter could make use of your census.

5. If you can obtain permission to use the class registers, tabulate a frequency distribution of the ages of the pupils in your school as follows:

Age in years	10–	11–	12–	13–	14–	15–	16–	17–	18 and over
No. of pupils									

Draw the frequency polygon. You may find that it is approximately rectangular up to 16 years of age.

6. A company which manufactures tubes for television receivers conducted a test of a sample batch of 1000 tubes and recorded the number of faults in each tube in the following table:

No. of faults	0	1	2	3	4	5	6
Frequency	620	260	88	20	8	2	2

(London.)

Plot the histogram and you will find it is an example of what is called *the reverse-J shape.*

7. The following table gives the distribution of 1000 families according to the number of children:

No. of children in family	0	1	2	3	4	5	6	7
No. of families	25	306	402	200	53	8	4	2

(London.)

Draw the histogram and note that it is skewed positively.

14. Determination of the mode. The following table shows the frequency distribution of the marks of 800 candidates in an examination:

Marks	1–10	11–20	21–30	31–40	41–50
No. of candidates	10	40	80	140	170

Marks	51–60	61–70	71–80	81–90	91–100
No. of candidates	130	100	70	40	20

Construct the histogram and find the mode. (London.)

The histogram is shown in fig. 5. Note that the bases of the rectangles extend from $\frac{1}{2}$ to $10\frac{1}{2}$, $10\frac{1}{2}$ to $20\frac{1}{2}$, etc., so as to *enclose* the ranges 1–10,

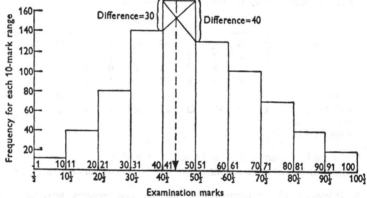

Fig. 5

11–20, etc. The modal class is 41–50. It is represented by the rectangle with base extending from $40\frac{1}{2}$ to $50\frac{1}{2}$. It contains 30 more candidates than the class below it and 40 more than the class above it. We therefore argue that the mode is *likely* to divide the modal class in the ratio 30:40. Hence we *estimate* the mode as

$$40\frac{1}{2} + \tfrac{3}{7} \text{ of } 10 \quad \text{or} \quad 50\frac{1}{2} - \tfrac{4}{7} \text{ of } 10,$$

and we write

$$\text{mode} = 44 \cdot 8,$$

which indicates that, if the marks are integers, the mode is likely to be 44 or 45.

Fig. 5 shows a geometrical method of determining the mode using the properties of similar triangles.

15. Exercises.

1. The following is the frequency distribution of the weights of 100 men:

Wt (lb.)	Frequency	Wt (lb.)	Frequency
100–109	1	160–169	17
110–119	2	170–179	10
120–129	5	180–189	6
130–139	11	190–199	4
140–149	21	200–209	2
150–159	20	210–219	1

Construct the histogram and find the mode. (London.)

2. In estimating the value of a plantation of fir trees, the girths of the trees in a sample area of 500 trees were measured and tabulated with a 10 in. grouping interval as follows:

Girth (in.)	15–25	25–35	35–45	45–55	55–65	65–75	75–85
No. of trees	25	30	135	160	100	40	10

Construct a histogram and estimate the mode. (London.)
Note this method of stating the groups 15–25, 25–35, etc. It is possible that there may be some trees whose girths are exactly 25 in. If so, half of them will have been placed in the 15–25 group and half in the 25–35 group.

16. Frequency distribution with unequal class intervals.
Let us now consider the age distribution of the male population of the United Kingdom in 1953. This is taken from *The Annual Abstract of Statistics* published by H.M. Stationery Office. The populations are given in thousands (table 1 C).

TABLE 1 C

Age group	Population	Age group	Population
Under 1	399	40–44	1890
1 and under 2	390	45–49	1833
2–4	1239	50–54	1593
5–9	2133	55–59	1286
10–14	1726	60–64	1081
15–19	1662	65–69	888
20–24	1701	70–74	669
25–29	1780	75–79	436
30–34	1900	80–84	206
35–39	1710	85 and over	76

The first group 'under 1' represents an interval of 1 year, the second group '1 and under 2' also represents an interval of 1 year, but '2–4' is 3 years while '5–9' is 5 years. If we divide these last two groups by 3 and 5 respectively we can rewrite the beginning of table 1 C as shown in table 1 D.

TABLE ID

Age group	Population (nearest integer)	Age group	Population (nearest integer)
Under 1	399	5 and under 6	427
1 and under 2	390	6 and under 7	427
2 and under 3	413	7 and under 8	427
3 and under 4	413	8 and under 9	427
4 and under 5	413	9 and under 10	427

We can now draw the histogram for this part of the age distribution as shown in fig. 6a. Note particularly that the scale 0, 100, 200, 300, 400 shows the population in thousands for *each 1-year class*.

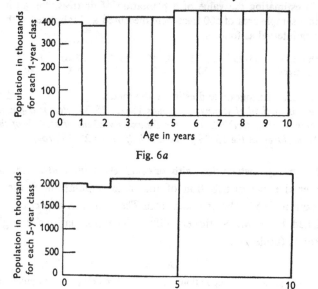

Fig. 6a

Fig. 6b

As the original age distribution is given mostly in 5-year intervals it would seem more satisfactory to convert fig. 6a into the form shown in fig. 6b.

The populations printed along the ordinate scale now are for *each 5-year* interval, and this fact is emphasized in the diagram by making the widths of the rectangles 5 years. The first three groups, 'under 1', '1 and under 2', '2–4', have been combined into one group, the total population of which can clearly be seen to be approximately 2,000,000, but the low population of the '1 and under 2' group is still represented in the diagram.

At the end of table 1 C we have the group '85 and over'. Before we can represent this diagrammatically we must first decide what interval of years it covers. Not many men live a hundred years but quite a number live more than 95 years, and hence it seems reasonable to regard this last group as '85 but under 100'. As this interval is 15 years its population will have to be divided by 3. This last group will, therefore, be represented by 3 rectangles each of width 5 years and height 25,000.

17. Comments on special features of the histogram. The complete age-distribution histogram is shown in fig. 7. From 40 years of age onwards it is exactly what we should expect. There is, however, an

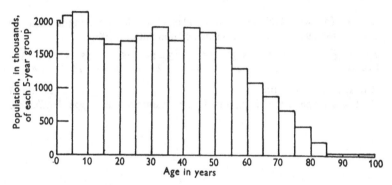

Fig. 7. Age distribution of the total male population of the United Kingdom in 1953, including members of H.M. Forces serving overseas but excluding Commonwealth and foreign forces in the United Kingdom.

unusually low population between 35 and 40 years. These men were born between 1913 and 1918, and during the 1914–18 war there was a low birth-rate. Moreover, in 1939 these same men were 21–26 years of age and many men of that age lost their lives serving in the 1939–45 war. The low population of the 15–20 years group is due to the low birth-rate between 1933 and 1938. The high population of the 5–10 years group is due to the high birth-rate immediately after the 1939–45 war. The drop in population of the '1 and under 2' group is due to the fact that in 1952, 29 babies out of every thousand died before they reached the age of 1 year. This *infant mortality* continues year by year, although, of course, the actual rate of 29 per thousand is not constant and hence, if the birth rate is constant the population of the '1 and under 2' group must always be less than that of the 'under 1' group.

18. Exercises.

1. The following table shows the analysis at September 1947 of grocers, provision merchants and general food shops according to the number of sugar registrations. Construct a histogram to illustrate the table.

No. of sugar registrations	No. of shops	
1–49	26,525	
50–99	29,420	
100–199	30,610	
200–299	15,349	
300–499	17,212	
500–999	17,014	
1000–1999	7,556	
2000 and over	2,362	(London.)

Note that it is skewed positively.

2. The following table gives the number of deaths occurring in each age group during 1952 for a certain village of population 6200:

Age group	0–5	5–15	15–25	25–35	35–45	45–55	55–65	Over 65
Deaths	13	1	2	3	3	8	12	20

Construct a histogram. (London.)

Note the shape of the histogram. It is an example of a *U-type distribution*.

2

Cumulative Frequency Distributions

19. The cumulative frequency distribution. The frequency distribution given in table 1 A can be converted to a *cumulative frequency distribution* by adding each frequency to the total of its predecessors. An appropriate way of tabulating it is given in table 2 A.

TABLE 2A

No. of tomatoes per truss	Cumulative frequency
Not more than 8	1
Not more than 9	2
Not more than 10	4
Not more than 11	9
Not more than 12	18
Not more than 13	33
Not more than 14	52
Not more than 15	72
Not more than 16	88
Not more than 17	98
Not more than 18	103
Not more than 19	106
Not more than 20	107
Not more than 21	107
Not more than 22	108

In table 2 A, for example, the cumulative frequency 18 for not more than 12 tomatoes per truss is the sum of the frequencies 1, 1, 2, 5, 9 given in table 1 A for 8, 9, 10, 11, 12 tomatoes per truss. It means that there are 18 of the 108 trusses with not more than 12 tomatoes. That is to say, one-sixth or, $16\frac{2}{3}$ %, of the trusses hold 12 tomatoes or less. It can also be seen that 72 of the 108 trusses have not more than 15 tomatoes. Thus it might be stated that two-thirds, or $66\frac{2}{3}$ %, of the trusses hold 15 tomatoes or less and this means, of course, that $33\frac{1}{3}$ % of the trusses hold 16 tomatoes or more.

20. The cumulative frequency curve or ogive. The graph of the cumulative frequency distribution shown in fig. 8 is called the *cumulative frequency curve* or *ogive*. The word 'ogive' is a term used in architecture to describe curves shaped like that in fig. 8.

21. Correct wording and correct plotting. If the frequency distribution given in table 1B is converted to a cumulative frequency distribution it can be set down as shown in table 2B.

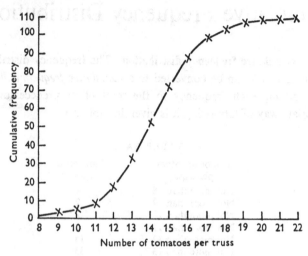

Fig. 8. Cumulative frequency curve or ogive.

TABLE 2B

Quantity of potatoes per root in lb.	Cumulative frequency
Under 3	9
Under 6	31
Under 9	59
Under 12	80
Under 15	97
Under 18	100

Attention is drawn to the correct phraseology in this case. *Under* 3 or *less than* 3 is correct but *not more than* 3 is wrong because 3 belongs to the second class in the original frequency distribution. *The method of stating the class intervals in the frequency distribution must always be carefully examined in order to decide whether 'less than' or 'not more than' is correct for the cumulative frequency distribution.* The actual drawing of the ogive of table 2B is left as an exercise for the student. It is important to plot the ordinates 9, 31, 59, etc., with abscissae 3, 6, 9, etc., respectively. A common mistake is to plot them with abscissae $1\frac{1}{2}$, $4\frac{1}{2}$, $7\frac{1}{2}$, etc., like the frequency polygon. *Abscissae of $1\frac{1}{2}$, $4\frac{1}{2}$, etc., are correct for the frequency polygon, but abscissae of 3, 6, etc., are correct for the ogive.*

22. Percentiles. The cumulative distribution of the examination marks given in §14 is shown in table 2C.

TABLE 2C

Examination mark	No. of candidates (cumulative frequency)
Not more than 10	10
Not more than 20	50
Not more than 30	130
Not more than 40	270
Not more than 50	440
Not more than 60	570
Not more than 70	670
Not more than 80	740
Not more than 90	780
Not more than 100	800

The student should plot an accurate ogive of table 2C in order to verify the answers to the following questions.

Question (i). *If the pass mark is 40, what percentage of the candidates pass the examination?*

The ogive shows that the number of candidates with not more than 39 is 256. Therefore 256 out of 800 fail. Hence 32 % fail and 68 % pass. Note that the examination mark 39 is called the *32nd percentile* because 32 % of the candidates obtain 39 or less.

Question (ii). *If it is decided to allow 80 % of the candidates to pass, what should the pass mark be?*

The ogive shows that 20 %, i.e. 160, of the candidates have not more than 32 marks. Hence the pass mark should be 33. In this case 32 is the 20th percentile.

Question (iii). *If scholarships are awarded to the top 15 % of the candidates, what would be the lowest mark to gain a scholarship?*

The ogive shows that 85 %, i.e. 680, of the candidates have not more than 71 marks. Hence 72 is the lowest mark to gain a scholarship. In this example 71 is the 85th percentile.

If an ogive is drawn from a cumulative frequency distribution in which the phrase 'less than' is used instead of 'not more than' the arguments used in (i), (ii) and (iii) above must be modified. For example, table 2B shows that 59 % of the potatoes have under 9 lb. of potatoes per root and hence 41 % have 9 lb. or more per root. The 59th percentile is still said to be 9 lb. per root.

23. The median. If a group of boys is arranged in order of height, the height of the middle boy is called the *median*. If, for example, the heights in inches of 7 boys are

$$61, \quad 63, \quad 63, \quad 64, \quad 68, \quad 71, \quad 71,$$

the height of the 4th boy, 64 in., is the median.

Similarly, if the ages of 5 boys are

14 yr. 3 mth.,   14 yr. 9 mth.,   14 yr. 9 mth.,
14 yr. 10 mth.,   14 yr. 11 mth.,

the age of the middle boy, i.e. the 3rd boy, 14 yr. 9 mth., is the median.

Further, if the weights in lb. of 10 men are

$$140, \quad 144, \quad 148, \quad 150, \quad 151, \quad 157, \quad 160, \quad 163, \quad 164, \quad 165,$$

the median is taken as $\frac{1}{2}(151+157)=154$ lb. because, due to the number of men in the group being even, the 5th and 6th men of weights 151 and 157 lb. are both in the middle. Thus, in this case, the median is defined as the arithmetic mean or average of the weights of the 5th and 6th men, and it is known as the weight of the 5·5th man. Hence it may be stated in general that *if N measurements or observations are arranged in order of magnitude, the* MEDIAN *is the $\frac{1}{2}(N+1)$th measurement or observation.*

If the populations of the 80 towns or villages given in §13, exercise 1, are arranged in order the median will be the population of the $\frac{1}{2}(80+1)$th = the 40·5 town or village. It will be found that the 40th and 41st towns or villages have both populations of 30 hundred. Hence the median is 3000.

24. The median of a large population. The marks of the 800 candidates given in table 2 C are an example of a *large population*. The median is the mark of the 400·5th candidate. The separate marks, however, are not available. Instead, the cumulative distribution is given and the ogive shows that 400·5 candidates get not more than 48 marks. It is fairly safe to assume, therefore, that the 400·5th candidate gets 48 marks and hence 48 is taken as the median. It will be realized that the difference between the 400·5th and the 400th is negligible. It cannot in fact be detected on the graph. Hence it should be noted that when the number of measurements N is large the median is taken as the $\frac{1}{2}N$th measurement. Alternatively, the median may be defined as the 50th percentile. It has half of the population above it and half below it. *It divides the area of the histogram into two equal parts.*

25. The quartiles. In a large population like that under consideration the *25th percentile* is called the *lower quartile* and the *75th percentile* the *upper quartile.* The ogive shows that 200 candidates have not more than 35 marks, while 600 candidates have not more than 63 marks.

Hence *the 25th percentile or lower quartile is 35 marks* and *the 75th percentile or upper quartile is 63 marks.*

The precise definition of the lower and upper quartiles is as follows: *If N measurements are arranged in order of magnitude, the $\frac{1}{4}(N+1)th$ and the $\frac{3}{4}(N+1)th$ are the LOWER QUARTILE and UPPER QUARTILE respectively.* It will be understood that $\frac{1}{4}(N+1)$th and $\frac{3}{4}(N+1)$th approximate to $\frac{1}{4}N$th and $\frac{3}{4}N$th respectively when N is large. Further, it will be realized that the lower quartile has one-quarter of the total population below it and three-quarters above it, while the upper quartile has three-quarters of the total population below it and one-quarter above it. *The median and quartiles together divide the area of the histogram into four equal parts.*

26. Exercises.

1. Arrange the examination marks of the 85 boys given in § 13, exercise 3, in order of magnitude and hence find the median and quartiles.

2. Use table 1 C to form a cumulative frequency distribution of the male population of the United Kingdom in 1953. Under 1, under 2, under 5, under 10, ..., under 100, would be the correct wording.
Draw the cumulative frequency curve or ogive and use it to estimate the percentage of the total male population (i) under 16, (ii) over 65, (iii) over 18 but under 20.

3. Form a cumulative frequency distribution of the girths of trees given in § 15, exercise 2. The appropriate wording in this case is not more than 25, not more than 35, etc.
Draw the ogive and use it to determine the median and quartiles.

4. Form a cumulative frequency distribution from the data of § 15, exercise 1. The wording suggested in this case is less than 110, less than 120, etc., because, either the weights have been given to the nearest lb., or fractions of a lb. have been neglected.
Draw the ogive and determine the median and quartiles.

5. Distinguish between *median* and *mode*.
In a savings group there are 800 members, and the numbers of savings certificates held by them are shown in the following frequency table.

Construct a table showing the cumulative frequency distribution and draw the graph of the ogive. Find (i) the mode, (ii) the median and (iii) the quartiles.

No. of certificates held	1–50	51–100	101–150	151–200	201–250
No. of members	5	10	30	60	100
No. of certificates held	251–300	301–350	351–400	401–450	451–500
No. of members	140	220	150	50	35

(London.)

6. The following table gives the frequency distribution of marks obtained by 130 candidates in each of two subjects A and B. Construct a table showing the cumulative frequency distributions in each subject and draw in one diagram the graphs of their ogives.

From your diagram, find

(a) the percentage number of candidates that fail in each subject if the pass mark in subject A is 55 and that in subject B is 35;

(b) the median mark in each subject.

Percentage	1–10	11–20	21–30	31–40	41–50
Subject A	0	0	1	3	6
Subject B	5	26	30	28	25
Percentage	51–60	61–70	71–80	81–90	91–100
Subject A	24	30	31	22	13
Subject B	9	5	0	1	1

(London.)

7. The following table gives the frequency distribution of the marks obtained in a test given to 300 candidates. Construct a table showing the cumulative frequency and draw a graph of the ogive. Obtain from the graph the median and the quartiles.

By extending the division of the distribution into percentiles, obtain also the mark not reached by the lowest 40 % of the candidates and the percentage passing if the pass mark is 50.

0–10	11–20	21–30	31–40	41–50	51–60
10	10	20	20	25	25
61–70	71–80	81–90	91–100	101–110	111–120
40	55	45	20	20	10

(London.)

8. The table below shows the frequency distribution of marks of 800 candidates at an examination. Construct a table showing the cumulative frequency distribution and draw a graph of the ogive.

Use your ogive to determine:

(i) the median mark;

(ii) the percentage number of candidates that fails if the pass mark is 50;

(iii) the quartiles.

1–10	11–20	21–30	31–40	41–50	51–60	61–70	71–80	81–90	91–100
30	50	100	150	150	130	90	60	30	10

(London.)

CUMULATIVE FREQUENCY

9. Represent the following distribution in the form of an ogive or cumulative frequency curve and hence obtain estimates of the median and quartiles:

Daily wages (shillings)	No. of earners within given wage limits
10–	1
20–	7
30–	24
40–	36
50–	25
60–	6
70–	1
Total	100

(London.)

10. Construct an ogive or cumulative frequency curve for the following distribution and read off the median and quartiles:

Gross Margins of 180 Department Stores, 1927
(net sales over $1,000,000)

Gross margins as percentage of net sales	Frequency
18·5–	1
20·5–	1
22·5–	3
24·5–	1
26·5–	10
28·5–	21
30·5–	31
32·5–	61
34·5–	36
36·5–	10
38·5–	5

(London.)

11. *Chest Girth and Height of Males aged 20 examined by Medical Boards, Great Britain, 1939*

Chest girth (in.)	% cases	Height (in.)	% cases
Under 31	0·9	Under 61	0·7
31–	6·6	61–	3·4
33–	28·5	63–	12·4
35–	40·0	65–	26·2
37–	19·2	67–	29·6
39–	3·9	69–	19·0
41 and over	0·9	71–	6·8
Total	100·0	73 and over	1·9
		Total	100·0

Draw the ogives for each of the above distributions.

Note that, due to the frequencies being stated as percentages, the ogives are graphs of percentiles.

Determine the median and quartiles of each distribution. (London.)

12. From the following data calculate a table showing the numbers of people below given limits of age, and draw a cumulative frequency diagram or ogive.

Age (years)	Estimated population of Bristol in 1937 (hundreds)	Age (years)	Estimated population of Bristol in 1937 (hundreds)
0–5	293	50–55	249
5–10	283	55–60	234
10–15	329	60–65	194
15–20	316	65–70	146
20–25	341	70–75	104
25–30	356	75–80	61
30–35	338	80–85	28
35–40	309	85–90	8
40–45	293	90–95	1
45–50	267	Total	4150

Neglecting any changes due to deaths or to children leaving or entering the Bristol district, estimate how many schoolchildren between the ages of 5 and 15 (i.e. over 5 but under 16) there will be in Bristol in 1942. Express this number as a percentage of the number between those limits of age in 1937. (Northern.)

3

Averages

27. The arithmetic mean of a set of numbers. The *arithmetic mean*, or more simply the *mean*, of the five numbers 23·46, 23·56, 23·67, 23·98, 22·43 is

$$\tfrac{1}{5}(23{\cdot}46 + 23{\cdot}56 + 23{\cdot}67 + 23{\cdot}98 + 22{\cdot}43) = \tfrac{1}{5}(117{\cdot}10) = 23{\cdot}42.$$

This number 23·42 is commonly called the *average* of the five numbers, but in the study of Statistics it is always called the *arithmetic mean* or *mean* to distinguish it from the other forms of average dealt with in this chapter.

The above calculation can be simplified in the following way. First subtract 23 from each of the numbers and write them as 0·46, 0·56, 0·67, 0·98, −0·57. Next calculate their mean 0·42. Finally, replace the 23 and thus obtain 23·42. This process is described as 'working with 23 as origin', the number 23 being called the 'arbitrary origin', 'starting point' or 'working zero'.

An even greater simplification is to write the numbers as 46, 56, 67, 98, −57, ignoring for the time being the fact that they are *hundredths*. Their mean is calculated as 42. The 42 is then written in hundredths as 0·42 and finally the 23 is replaced giving 23·42. This is described as 'working with 23 as origin, and 0·01 as unit'.

Second example. By working with 15 years as origin and 1 month as unit, calculate, to the nearest month, the mean age of eight boys whose ages are 15 yr. 3 mth., 15 yr. 9 mth., 14 yr. 10 mth., 15 yr. 0 mth., 15 yr. 7 mth., 14 yr. 4 mth., 15 yr. 5 mth., 16 yr. 0 mth.

Calculation. The mean of 3, 9, −2, 0, 7, −8, 5, 12 is

$$\tfrac{1}{8}(3 + 9 - 2 + 0 + 7 - 8 + 5 + 12) = 3\tfrac{1}{4} \text{ months,}$$

and hence the mean age = 15 yr. 3¼ mth.

$$= 15 \text{ yr. } 3 \text{ mth. (to the nearest month).}$$

28. Exercises.

1. Working with 980 as origin and 0·1 as unit calculate the mean of 980·8, 981·1, 980·7, 980·3, 981·8, 982·5.

2. A cricketer makes the following scores in ten completed innings: 47, 41, 50, 39, 45, 48, 42, 32, 60, 20. Calculate his batting average by taking 40 as origin.

3. Calculate, correct to 1 decimal place, the mean of the following percentage relatives: 100·2, 101·3, 101·8, 98·4, 100·5, 100·2, 101·6, 97·5, 99·8.

29. When some of the observations are repeated. If the heights, in inches, of twelve men are given as 73, 72, 72, 71, 71, 71, 70, 70, 69, 69, 69, 69, the mean height is

$$70 + \tfrac{1}{12}(3 + 2 + 2 + 1 + 1 + 1 + 0 + 0 - 1 - 1 - 1 - 1) = 70\tfrac{1}{2} \text{ in.}$$

The heights might have been given in a frequency table:

Height (in.)	73	72	71	70	69
Frequency	1	2	3	2	4

The mean height might then have been calculated as

$$70 + \tfrac{1}{12}\{3 + 2(2) + 3(1) + 2(0) + 4(-1)\} = 70\tfrac{1}{2} \text{ in.}$$

It is convenient to tabulate the last calculation as given in table 3A.

TABLE 3A

Height in inches taking 70 as origin x	Frequency f	Frequency × height fx
3	1	3
2	2	4
1	3	3
0	2	0
−1	4	−4
Total	12	6

$$\text{Mean height} = 70 + \left(\frac{\text{Sum of the } fx \text{ column}}{\text{Sum of the } f \text{ column}}\right) \text{ in.}$$

$$= 70\tfrac{1}{2} \text{ in.}$$

30. The mean of a frequency distribution. The method of the last paragraph can be applied to the frequency distribution in table 1A to calculate the mean number of tomatoes per truss. In this case it is convenient to take the mode 15 tomatoes per truss as origin, so that when f is large, x is small, and the product fx is easily calculated as shown in table 3B.

TABLE 3B

No. of tomatoes per truss	Frequency f	No. of tomatoes per truss taking 15 as origin x	fx	
8	1	−7		− 7
9	1	−6		− 6
10	2	−5		−10
11	5	−4		−20
12	9	−3		−27
13	15	−2		−30
14	19	−1		−19
15	20	0	0	
16	16	1	16	
17	10	2	20	
18	5	3	15	
19	3	4	12	
20	1	5	5	
21	0	6	0	
22	1	7	7	
Total	108	Total	75	−119

In the fx column, positive products are placed on the left and negative products on the right so that they can be summed separately.

$$\text{Mean} = 15 + \left(\frac{\text{Sum of the } fx \text{ column}}{\text{Sum of the } f \text{ column}}\right)$$

$$= 15 + \left(\frac{75 - 119}{108}\right)$$

$$= 14 \cdot 6 \text{ tomatoes per truss.}$$

31. Exercises.

1. Rewrite the calculation shown in §30 taking 14 as origin instead of 15 and verify that an identical result is obtained.

2. Use the data of §13, exercise 6 to calculate the mean number of faults per television tube. As the mode is 0 it is not necessary to change the origin.

3. Use the data of §13, exercise 7 to calculate the mean number of children per family. Take 2 children in family as origin.

32. The use of mid-interval values. To calculate the mean weight of potatoes per root from the data of table 1 B it is customary to take the 9 roots yielding under 3 lb. per root as each yielding $1\frac{1}{2}$ lb. per root, their total yield being $13\frac{1}{2}$ lb. Similarly, the 22 roots yielding 3–6 lb. per root are taken as having a total yield of $22 \times 4\frac{1}{2}$ lb. This assumption is, of course, not altogether justified, but the larger the frequency of each group the nearer it becomes to being true. If the *mid-interval weight* $7\frac{1}{2}$ lb. of the modal class is taken as the arbitrary origin or working zero and 3 lb. are taken as one unit, the calculation is as shown in table 3C.

TABLE 3C

Weight of potatoes per root in lb.	Frequency f	Mid-interval weights in lb. of the groups in the first column	Mid-interval weights, $7\frac{1}{2}$ lb. as origin 3 lb. as unit x		fx	
Under 3	9	$1\frac{1}{2}$	-2			-18
3–	22	$4\frac{1}{2}$	-1			-22
6–	28	$7\frac{1}{2}$	0	0		
9–	21	$10\frac{1}{2}$	1	21		
12–	17	$13\frac{1}{2}$	2	34		
15–18	3	$16\frac{1}{2}$	3	9		
Total	100	—	Total	64	-40	

As x is measured in 3lb. units, fx is in 3lb. units and so it is necessary to multiply $\left(\dfrac{\text{Sum of the } fx \text{ column}}{\text{Sum of the } f \text{ column}}\right)$ by 3 to bring it to 1lb. units before adding it to the $7\frac{1}{2}$lb. which was taken as origin. Hence the mean is

$$7\frac{1}{2} + \frac{3(64-40)}{100} = 8\cdot22 \text{ lb. per root.}$$

Note that an alternative way of describing the values x is, 'deviations of the mid-interval weights from $7\frac{1}{2}$ lb. taking 3 lb. as unit', since, for example, $13\frac{1}{2}$ lb. deviates from $7\frac{1}{2}$ lb. by $+2$ 3 lb. units while $1\frac{1}{2}$ lb. deviates from $7\frac{1}{2}$ lb. by -2 3 lb. units.

33. The use of Σ. The phrase 'the *sum* of the fx column' is used so often in statistical calculations that it is convenient to represent it by Σfx. The symbol Σ, pronounced 'sigma', is the capital S of the Greek alphabet. Similarly Σf is short for 'the sum of the f column'.

34. A second example using mid-interval values. The method of §32 can be applied to the data of §14 to calculate the mean examination mark of the 800 candidates as shown in table 3D.

Note that the way in which the groups of a frequency distribution are stated must always be carefully studied to decide the mid-interval values. An error in the mid-interval values causes an equal error in the mean Examination marks grouped 0–9, 10–19, etc., have mid-interval values $4\frac{1}{2}$, $14\frac{1}{2}$, etc., but weights of men in lb. grouped 120–, 130–, etc., would have mid-interval weights 125, 135, etc., because '120–' means

120–130 but not including 130. The variation in weights is *continuous* with the dividing line drawn at 130, whereas the variation in the marks is *discrete* with 9 belonging to one group and 10 to the next, there being no possible mark between.

TABLE 3D

Examination mark	No. of candidates f	Mid-interval marks of the groups of the first column	Mid-interval marks $45\frac{1}{2}$ as origin 10 as unit x		fx
1–10	10	$5\frac{1}{2}$	−4		− 40
11–20	40	$15\frac{1}{2}$	−3		−120
21–30	80	$25\frac{1}{2}$	−2		−160
31–40	140	$35\frac{1}{2}$	−1		−140
41–50	170	$45\frac{1}{2}$	0	0	
51–60	130	$55\frac{1}{2}$	1	130	
61–70	100	$65\frac{1}{2}$	2	200	
71–80	70	$75\frac{1}{2}$	3	210	
81–90	40	$85\frac{1}{2}$	4	160	
91–100	20	$95\frac{1}{2}$	5	100	
Total	800	—	Total	800	−460

$$\text{Mean examination mark} = 45\tfrac{1}{2} + \frac{10\Sigma fx}{\Sigma f}$$

$$= 45\tfrac{1}{2} + \frac{10(800-460)}{800}$$

$$= 49\tfrac{3}{4}.$$

35. Exercises.

1. Rewrite the calculation shown in §32 taking $10\frac{1}{2}$ lb. as origin and 3 lb. as unit and verify that an identical result is obtained.

2. Rewrite the calculation shown in §34 taking $55\frac{1}{2}$ as origin and 10 as unit and verify that an identical result is obtained.

3. Taking a working zero of $154\frac{1}{2}$ lb. and 10 lb. as unit, calculate the mean weight of the 100 men given in §15, exercise 1.

4. By taking an arbitrary origin in the 45–55 group calculate the mean girth of the trees given in §15, exercise 2.

5. The 'recovery time' of an aircraft is the time that elapses between its arrival at an airport and its being ready to take-off again.
Calculate the mean recovery time of the 100 aircraft shown in the following distribution:

Recovery time in minutes	5–10	10–15	15–20	20–25
No. of aircraft	4	22	28	18
Recovery time in minutes	25–30	30–35	35–40	40–45
No. of aircraft	12	9	4	3

6. Use the data of §26, exercise 5, to obtain the mean holding of certificates.

7. Use the data of §26, exercise 9, to obtain the mean daily wage.

8. Use the data of §26, exercise 11, to calculate the mean chest girth and height.

36. Unequal grouping. The calculation of the mean age of the male population of the United Kingdom in 1953 from the data of table 1C involves dealing with unequal class intervals. The mid-interval ages for under 1, 1 and under 2, and 85 and over are respectively $\frac{1}{2}$, $1\frac{1}{2}$ and $92\frac{1}{2}$, 85 and over being regarded as 85–99. The mid-interval of the 2–4 group is $3\frac{1}{2}$ because this group may include boys who are only 1 day short of being 5. Similarly, the mid-interval of the 5–9 groups is $7\frac{1}{2}$. The full calculation is shown in table 3E.

TABLE 3E

Mid-interval age in years	Population (thousands) f	Mid-interval age $37\frac{1}{2}$ as origin 5 years as unit x	fx
$\frac{1}{2}$	399	−7·4	− 2,952·6
$1\frac{1}{2}$	390	−7·2	− 2,808·0
$3\frac{1}{2}$	1,239	−6·8	− 8,425·2
$7\frac{1}{2}$	2,133	−6	−12,788
$12\frac{1}{2}$	1,726	−5	− 8,630
$17\frac{1}{2}$	1,662	−4	− 6,648
$22\frac{1}{2}$	1,701	−3	− 5,103
$27\frac{1}{2}$	1,780	−2	− 3,560
$32\frac{1}{2}$	1,900	−1	− 1,900
$37\frac{1}{2}$	1,710	0	0
$42\frac{1}{2}$	1,890	1	1,890
$47\frac{1}{2}$	1,833	2	3,666
$52\frac{1}{2}$	1,593	3	4,779
$57\frac{1}{2}$	1,286	4	5,144
$62\frac{1}{2}$	1,081	5	5,405
$67\frac{1}{2}$	888	6	5,328
$72\frac{1}{2}$	669	7	4,683
$77\frac{1}{2}$	436	8	3,488
$82\frac{1}{2}$	206	9	1,854
$92\frac{1}{2}$	76	11	836
Total	24,598	Total	37,073 −52,824·8

$$\text{Mean age} = 37\tfrac{1}{2} + \frac{5(-15751\cdot8)}{24598}$$

$$= 34\cdot3 \text{ years.}$$

37. Discussion of averages in general. The *arithmetic mean, mode* and *median* are all averages. Two other forms of average, the *geometric mean* and *harmonic mean*, will also be described and discussed in this chapter. An average can be described as a numerical characteristic of a group which enables us to assess the position in which the group stands with respect to other groups. The group under consideration, for example, might be a variety of potato whose position with respect to other varieties is being assessed by the average weight of potatoes its produces per root, or it might be a workshop whose position with respect to other workshops is being assessed by the average number of articles it produces per week. It might, indeed, be an individual such as a cow whose position in the herd is assessed by its average milk yield, or, of course, a cricketer whose position among other cricketers is assessed by his batting or bowling average.

38. The arithmetic mean. The arithmetic mean is the best known and most useful form of average. It is the basis of all statistical theory. By multiplying the mean by the total frequency the aggregate for the group is obtained. Means of separate groups can easily be manipulated to obtain the mean of the combined group. For example, if the mean age of 15 boys is m_1 years and the mean age of 14 girls is m_2 years, the mean age of the 29 boys and girls is $\frac{1}{29}(15m_1 + 14m_2)$ years. Sometimes abnormal individuals among the observations can have a rather exaggerated effect on the mean, and this is the chief criticism that is usually levelled against it. Suppose, for example, the National Savings Collection at a school was approximately £5 each week for 40 weeks but £210 on one particular week when a special campaign was organized. The mean weekly collection $= £\frac{1}{41}(40 \times 5 + 210) = £10$. Thus the one abnormal individual has made the mean equal to double the normal collection.

39. The median. The median can easily be picked out when the individuals are ranged in order, and it provides an *average specimen* for examination. In a form, the boy of median age can be interviewed, the physique of the boy of median weight can be examined, and the exercise books of the boy with the median mark can be studied carefully. The median is entirely unaffected by abnormal individuals, but it is unsuitable for work demanding mathematical manipulation.

40. The mode. The mode seems most useful to manufacturers of articles by sizes. It is obviously useful for the shoemaker to know the modal size of shoe and for the cycle manufacturer to know the modal size of frame. The mode is not easy to determine with precision when the observations fall into groups. Indeed, if the variation is continuous it is quite possible that no two observations may be alike. This was mentioned in §6. Like the median, the mode is unaffected by abnormal individuals and is unsuitable for mathematical manipulation.

41. Relation between mean, median and mode. An interesting method of estimating approximately the mode is by the formula

$$\text{mode} = \text{mean} - 3(\text{mean} - \text{median}).$$

In §§24 and 34, for example, we found that the median and mean marks of the 800 candidates were 48 and $49\frac{3}{4}$ respectively. Hence the mode = $49\frac{3}{4} - 3(49\frac{3}{4} - 48) = 44\frac{1}{2}$, which is approximately the estimation of §14.

*42. The geometric mean. The *geometric mean* of *n* numbers is the *n*th root of their product. Thus the geometric mean of the five numbers 1, 2, 3, 4, 5 is $\sqrt[5]{(1.2.3.4.5)} = 2\cdot605$ by logarithms. Note that the geometric mean is less than the arithmetic mean. It is not influenced by abnormal individuals to the same extent. Further, the geometric mean of two ratios p/a and q/b is equal to the ratio of the geometric mean $\sqrt{(pq)}$ to the geometric mean $\sqrt{(ab)}$. This can easily be verified for any four positive numbers a, b, p, q. This property, which does not hold for the arithmetic mean, makes the geometric mean more satisfactory for averaging percentage relatives (see chapter 9).

*43. The harmonic mean. The *reciprocal* of the *harmonic mean* of *n* numbers is the *arithmetic mean of their reciprocals*. Thus the harmonic means of 1, 2, 3, 4, 5 is given by

$$\frac{1}{\text{harmonic mean}} = \frac{1}{5}\left(\frac{1}{1} + \frac{1}{2} + \frac{1}{3} + \frac{1}{4} + \frac{1}{5}\right)$$
$$= \tfrac{1}{5}(1 + 0\cdot5 + 0\cdot3333 + 0\cdot25 + 0\cdot2)$$
$$= 0\cdot4567.$$

Hence the harmonic mean = $2\cdot19$.

It should be noted that the harmonic mean is less than the geometric mean and that it might be alternatively defined as the *reciprocal of the arithmetic mean of the reciprocals*.

---

* The five sections (42, 43, 57, 61 and 62) marked with an asterisk are more difficult than the rest of the book and may be omitted at the first reading.

AVERAGES

The use of the harmonic mean is illustrated by the following example. If a boat travels s miles upstream at u m.p.h. and then returns the s miles downstream at v m.p.h., the $2s$ miles have been covered in $\left(\dfrac{s}{u}+\dfrac{s}{v}\right)$ hours, and hence the average speed for the whole journey is

$$\frac{2s}{\dfrac{s}{u}+\dfrac{s}{v}}=\frac{1}{\dfrac{1}{2}\left(\dfrac{1}{u}+\dfrac{1}{v}\right)}$$ m.p.h., which is the harmonic mean of u and v.

44. Exercises.

1. An aircraft travels 500 miles at 200 m.p.h. and returns over the same route at 250 m.p.h. Calculate its average speed for the whole journey.

2. A car does its first 5000 miles at 40 miles to the gallon of petrol and its next 5000 miles at 36 miles to the gallon. What is its average petrol consumption for the whole 10,000 miles?

4

Dispersion

45. Dispersion or variability. Suppose that the runs scored by two batsmen in seven completed innings are as shown in the following table:

	Scores for seven completed innings							Mean score
Batsman A	20	65	0	50	100	35	80	50
Batsman B	40	55	35	50	65	45	60	50

It is clear that, although the mean score of each batsman is the same, the scores of A are much more dispersed, scattered or spread about the mean than those of B. The scores of 100 and 80 made by A are better than any made by B, the scores of 0 and 20 are worse. The performances of A are more variable than those of B, and indeed, less reliable. The purpose of this chapter is to investigate the methods by which the *dispersion* or *variability* of groups of observations can be compared.

46. Examples.

1. A well-known sugar refiner uses machines which pack automatically 1 lb. cartons of sugar. When these machines are in operation the cartons emerge from them in a continuous stream ready for sale. In order to check that a particular machine is giving correct weight sample cartons are occasionally selected from it and weighed accurately. The results of checking two machines are as follows:

	Accurate weights in ounces of eleven 1 lb. cartons					
Machine A	16·17	16·51	16·78	15·96	16·33	16·59
Machine B	15·95	16·09	16·28	16·36	16·00	16·17
Machine A	16·82	16·14	16·40	16·72	15·98	
Machine B	16·27	16·45	16·06	16·18	16·39	

Note that the dispersion of A is greater than that of B because the weights delivered by machine A *range* from 15·96 to 16·82 oz., while those delivered by machine B *range* from 15·95 to 16·45 oz.

2. A firm which manufactures lead-covered submarine cables checked the thickness of the cover on two of its cables by taking measurements in ten places:

Ten measurements, in inches, of the thickness of the cover

Cable A	0·24	0·26	0·28	0·20	0·22	0·23	0·25	0·27	0·29	0·21
Cable B	0·20	0·22	0·23	0·24	0·22	0·22	0·24	0·21	0·22	0·23

Here the measurements *A range* from 0·20 to 0·29 while measurements *B range* from 0·20 to 0·24, and hence the dispersion of *A* is greater than that of *B*.

3. *English and Mathematics marks. Frequency distribution of 1000 candidates*

Examination mark	0–9	10–19	20–29	30–39	40–49
No. of candidates in English	1	3	19	131	346
No. of candidates in Mathematics	20	41	98	149	192

Examination mark	50–59	60–69	70–79	80–89	90–99
No. of candidates in English	340	134	20	4	2
No. of candidates in Mathematics	197	152	97	44	10

Since the number of candidates with very low and very high marks is greater in Mathematics than in English, the dispersion of the Mathematics marks may be taken to be greater than the dispersion of the English marks. If the frequency polygons of the two distributions are drawn they show quite clearly the difference in dispersion. In fig. 9 the

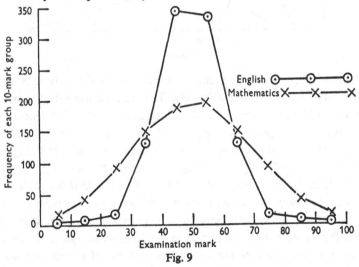

Fig. 9

tall narrow frequency polygon for the English marks indicates that a great number of candidates are packed close to the mean, while the short wide frequency polygon for the Mathematics marks indicates that the candidates are widely scattered away from the mean. Hence the dispersion is small for English and large for Mathematics.

4. A study of the weights of 100 cats and 100 dogs led to the following statements:

(i) The average weight of the dogs was greater than that of the cats.

(ii) The variability in weight was greater for the dogs.

(iii) The smallest dog was about the same weight as the smallest cat.

(iv) The biggest cats were smaller than the average dog.

(v) The distribution of the weights of the cats was approximately normal.

(vi) The distribution of the weights of the dogs was skewed positively.

It is interesting to express the above statements diagrammatically by means of frequency curves as shown in fig. 10.

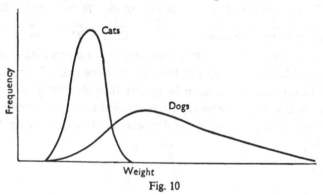

Fig. 10

47. The range. Having discussed dispersion qualitatively it is now necessary to assess it quantitatively. The simplest numerical measure of dispersion is the *range*. Thus the range of the scores of batsmen A in §45 is 100 runs (since his highest and lowest scores are 100 and 0), while the range of batsmen B is 30. In §46, example 1, the range of machine A is 0·86 oz., while that of machine B is 0·50 oz. In §46, example 2, the range for cable A is 0·09 in., while that for cable B is 0·04 in. The range, however, is not a good measure of dispersion. It is based entirely on the two extreme observations of a group. It is possible for the ranges of two groups to be the same and yet their dispersions may be different. The examination marks of eleven candidates in Art, for example, might be

0, 25, 35, 40, 45, 50, 55, 60, 65, 75, 100,

while their marks in French might be

0, 10, 20, 30, 40, 50, 60, 70, 80, 90, 100.

Here, except for the 0 and the 100, the Art marks are clustered more closely to the mean than the French. Hence the dispersion of the Art marks is less than the dispersion of the French marks, although the range is the same for each. Moreover, the range generally depends on the size of the sample, i.e. on the number of observations. The range of scores in twenty innings, for example, is likely to be greater than the range for seven innings.

48. Interpercentile ranges. A second measure of dispersion is the range between the 10th percentile and the 90th percentile. If the ogives are drawn for §46, example 3, the following results will be obtained from them:

	English marks	Mathematics marks
90th percentile	63½	74
10th percentile	35	23
10–90 percentile range	28½	51

This indicates that the dispersion of the Mathematics marks is almost twice that of the English marks. The 10–90 *percentile range* does not depend on the extreme observations and it does make some use of the whole group of observations, but it does not lend itself to further mathematical treatment in statistical analysis.

49. The semi-interquartile range or quartile deviation. A more frequently used measure of dispersion than the 10–90 percentile range is the *semi-interquartile range*. This, as its name implies, is half the range between the quartiles. Quite often it is called the *semi-quartile range* or *quartile deviation*. Thus

Semi-interquartile range or quartile deviation

$$= \frac{\text{Upper quartile} - \text{Lower quartile}}{2}.$$

From the ogive of §46, example 3, the following results can be obtained:

	English marks	Mathematics marks
Upper quartile	56	62½
Lower quartile	42	35
Interquartile range	14	27½
Semi-interquartile range	7	13¾

Thus the dispersion of the Mathematics marks is again indicated as almost twice that of the English marks. As the semi-interquartile range ignores completely the top quarter and bottom quarter of the observations and is based entirely on the range of the middle half, it is not a satisfactory measure of the dispersion of the whole group. Nor does it lend itself to further mathematical treatment.

50. The mean deviation. A measure of dispersion which makes use of all the observations of a group is the *mean deviation*. This is calculated for the machine *A* of §46, example 1, in table 4A in which

(i) the first column contains the weights of the cartons together with their total and mean;

(ii) the second column contains the differences between the weights and their mean. These differences are called the *deviations from the mean*;

(iii) the third column contains these *deviations all taken as positive.*

TABLE 4A

	Weights of cartons (oz.)	Deviations of the weights from the mean 16·40 oz. d oz.	Deviations all taken as positive $\|d\|$ oz.
	16·17	−0·23	0·23
	16·51	+0·11	0·11
	16·78	+0·38	0·38
	15·96	−0·44	0·44
	16·33	−0·07	0·07
	16·59	+0·19	0·19
	16·82	+0·42	0·42
	16·14	−0·26	0·26
	16·40	0·00	0·00
	16·72	+0·32	0·32
	15·98	−0·42	0·42
Total	180·40	0·00	2·84
Mean	16·40	—	0·26

The mean deviation= the mean of the deviations all taken as positive
= 0·26 oz.

Note that $\|d\|$ means the magnitude of d without regard to its sign. It is called the *modulus* of d.

The student will find it a useful exercise to calculate by the same method the mean deviation for machine *B*. The result obtained, 0·14 oz., indicates that the dispersion or variability of machine *A* is almost twice that of machine *B*.

In calculating the mean deviation, the deviations may be measured from the MEAN, MEDIAN *or* MODE. *They are really best measured from the median, however, because the* MEAN DEVIATION FROM THE MEDIAN *is a* MINIMUM.

When the variability or dispersion of one group is being compared with that of another group the mean deviation must clearly be measured from the same average for each group.

51. Exercises.

1. Compare the variability in the thickness of the cover of the submarine cables of §46, example 2, by calculating the mean deviation from the median in each case. Note that the median for cable A is 0·245 in., while that for cable B is 0·22 in.

2. Calculate the mean deviation from the mean of the six values given in §28, exercise 1.

3. Calculate the mean deviation from the mean of the ten scores of §28, exercise 2.

4. Determine the median and the mean deviation from the median of the nine percentage relatives of §28, exercise 3.

5. The following are the index numbers of production in the textile and food industries for 10 years between 1924 and 1934, excluding 1926:

Textiles	Food
100	100
122	97
120	103
112	103
114	108
90	109
89	113
101	112
109	110
112	115

Compare the variability of production in the two industries by calculating the mean deviation from the median for each set of index numbers.

(Northern.)

6. Compare the variability in price of beef and mutton by calculating the mean deviation from the median for the prices shown in the following table:

Price in pence per 8 lb.

Year	Beef	Mutton
1925	80	106
1926	74	89
1927	70	86
1928	74	92
1929	71	89
1930	73	92
1931	67	79
1932	65	63
1933	61	69
1934	58	74
1935	54	75
1936	54	73
1937	61	78
1938	62	62

(Northern.)

52. Mean deviation of a frequency distribution. Table 4B shows the calculation of the mean deviation from the median of the English marks given in §46, example 3. The median is $49\frac{1}{2}$ because 500 candidates have 49 or less and 500 candidates have 50 or more.

TABLE 4B

Mid-interval mark	Frequency f	Deviations of mid-interval mark from the median $49\frac{1}{2}$ d	Deviations all taken as positive $\lvert d \rvert$	$f\lvert d \rvert$
$4\frac{1}{2}$	1	-45	45	45
$14\frac{1}{2}$	3	-35	35	105
$24\frac{1}{2}$	19	-25	25	475
$34\frac{1}{2}$	131	-15	15	1965
$44\frac{1}{2}$	346	-5	5	1730
$54\frac{1}{2}$	340	$+5$	5	1700
$64\frac{1}{2}$	134	$+15$	15	2010
$74\frac{1}{2}$	20	$+25$	25	500
$84\frac{1}{2}$	4	$+35$	35	140
$94\frac{1}{2}$	2	$+45$	45	90
Total	1000	—	—	8760

$$\text{Mean deviation} = \frac{\text{Sum of the } f\lvert d \rvert \text{ column}}{\text{Sum of the } f \text{ column}}$$

$$= \frac{\Sigma f\lvert d \rvert}{\Sigma f}$$

$$= 8\cdot76.$$

The student should calculate by the same method the mean deviation of the Mathematics marks. It is an unusual coincidence that the median is the same as for the English marks. The value obtained for the mean deviation should be 15·66, thus indicating that the dispersion of the Mathematics marks is almost twice that of the English marks.

53. Exercises.
Calculate the mean deviation from the median for the following:

1. The number of certificates of §26, exercise 5.
2. The subjects A and B of §26, exercise 6.
3. The examination marks of §26, exercise 7.
4. The daily wages of §26, exercise 9.
5. The gross margin percentages of §26, exercise 10.

54. The standard deviation. The most satisfactory and most widely used measure of dispersion is the *standard deviation*. This is calculated for a small number of observations as follows:

(i) Write down the *squares of the deviations* from the mean. These will, of course, all be positive.

(ii) Calculate the *mean of these squares of deviations*.

(iii) The standard deviation is the *square root* of *the mean of the squares of deviations*.

The standard deviation is always measured from the MEAN *and never from the median or mode.*

For the machine A of §46, example 1, the calculation of the standard deviation might be tabulated as shown in table 4C.

TABLE 4C

Weights of cartons (oz.)	Deviations of the weights from the mean 16·40 oz. d oz.	Squares of the deviations d^2 oz.^{2}
16·17	−0·23	0·0529
16·51	+0·11	0·0121
16·78	+0·38	0·1444
15·96	−0·44	0·1936
16·33	−0·07	0·0049
16·59	+0·19	0·0361
16·82	+0·42	0·1764
16·14	−0·26	0·0676
16·40	0·00	0·0000
16·72	+0·32	0·1024
15·98	−0·42	0·1764
Total 180·40	0·00	0·9668
Mean 16·40	—	0·0879

$$\text{Standard deviation} = \sqrt{\left(\frac{\Sigma d^2}{11}\right)} \text{ oz.}$$

$$= 0·296 \text{ oz.}$$

A similar calculation for machine B gives the standard deviation 0·157 oz., thus indicating that the dispersion of machine A is almost twice that of machine B.

55. Avoiding the decimals. If the deviations in the last calculation had been written with 0·01 oz. as unit the decimals would have been avoided. The total for the d^2 column would then be 9668, its mean 879 and the final step in the calculation would be

$$\text{Standard deviation} = 0·01 \sqrt{\left(\frac{\Sigma d^2}{11}\right)}$$

$$= 0·01 \times 29·6 \text{ oz.}$$

$$= 0·296 \text{ oz.}$$

The factor 0·01 is introduced to bring back the units to oz.

56. Using an arbitrary origin. Much arithmetical labour can be avoided by using an arbitrary origin in the calculation of the standard deviation. Consider the thicknesses of the cover of cable A given in §46, example 2. Their mean is 0·245 in., giving deviations $-0·005$, $+0·015$, $+0·035$, etc., and even if the decimals are avoided by working with 0·001 in. as unit the squares are still rather cumbersome. Table 4D shows the method of easing the burden by working with one of the central observations 0·25 in. as origin.

TABLE 4D

Thickness of cover in.	Thickness with 0·25 in. as origin and 0·01 in. as unit x	Squares of deviations from 0·25 in. with 0·01 in. as unit x^2
0·24	-1	1
0·26	$+1$	1
0·28	$+3$	9
0·20	-5	25
0·22	-3	9
0·23	-2	4
0·25	0	0
0·27	$+2$	4
0·29	$+4$	16
0·21	-4	16
Total	-5	85

By the method of §3·01

$$\text{the mean} = 0·25 + \frac{0·01 \times (-5)}{10} \text{ in.}$$

$$= 0·245 \text{ in.}$$

Now $\Sigma x^2 = 85$ is the sum of the squares of deviations from 0·25 in. and therefore $0·01 \sqrt{\left(\frac{\Sigma x^2}{10}\right)}$ is *not* the standard deviation. It can be proved, however, that

$$\text{the standard deviation} = 0·01 \sqrt{\left\{\frac{\Sigma x^2}{10} - \left(\frac{\Sigma x}{10}\right)^2\right\}},$$

the term $-\left(\frac{\Sigma x}{10}\right)^2$ being the correction necessary because the deviations are from 0·25 in. instead of 0·245 in. Hence

$$\text{the standard deviation} = 0·01 \sqrt{\{8·5 - (-0·5)^2\}} \text{ in.}$$

$$= 0·01 \sqrt{\{8·5 - 0·25\}} \text{ in.}$$

$$= 0·02872 \text{ in.}$$

It can be verified that, if deviations are measured from the true mean 0·245 in., the arithmetic is more awkward, the sum of the x column is zero, the sum of the x^2 column is 82·5 and hence the result obtained is the same.

The student will find it a useful exercise to calculate by the same method the standard deviation for cable B. By working with 0.22 in. as origin and 0.01 in. as unit it will be found that *zero* occurs four times in the x column. This makes the arithmetic very easy. The results obtained should be

$$\text{Mean} \qquad\qquad = 0.223 \text{ in.,}$$

$$\text{Standard deviation} = 0.0119 \text{ in.}$$

*57. Mathematical justification of the method. The method of §56 will now be justified by considering the 10 observations $x_1, x_2, x_3, ..., x_{10}$. Let $\bar{x}$ be their mean so that the deviation of x_1 from $\bar{x}$ is $x_1 - \bar{x}$. The following table shows the necessary steps in the calculation, from first principles, of the standard deviation:

10 observations x	Deviations of the observations from the mean $\bar{x}$ d	Squares of deviations d^2
x_1	$x_1 - \bar{x}$	$x_1^2 - 2x_1\bar{x} + \bar{x}^2$
x_2	$x_2 - \bar{x}$	$x_2^2 - 2x_2\bar{x} + \bar{x}^2$
x_3	$x_3 - \bar{x}$	$x_3^2 - 2x_3\bar{x} + \bar{x}^2$
$\vdots$	$\vdots$	$\vdots$
x_{10}	$x_{10} - \bar{x}$	$x_{10}^2 - 2x_{10}\bar{x} + \bar{x}^2$
Total $\quad \Sigma x$	Total	$\Sigma x^2 - 2\bar{x}\Sigma x + 10\bar{x}^2$

Note that the sum of the d^2 column, Σd^2, is

$$(x_1^2 + x_2^2 + x_3^2 + ... + x_{10}^2) - 2\bar{x}(x_1 + x_2 + x + ... + x_{10})$$
$$+ (\bar{x}^2 + \bar{x}^2 + \bar{x}^2 + ... + \bar{x}^2),$$

which can be written, as shown,

$$\Sigma x^2 - 2\bar{x}\Sigma x + 10\bar{x}^2.$$

Now the mean $\bar{x} = \dfrac{\Sigma x}{10}$ and, by definition, the standard deviation is

$$\sqrt{\left\{\frac{\Sigma d^2}{10}\right\}} = \sqrt{\left\{\frac{\Sigma x^2 - 2\bar{x}\Sigma x + 10\bar{x}^2}{10}\right\}}$$

$$= \sqrt{\left\{\frac{\Sigma x^2}{10} - 2\left(\frac{\Sigma x}{10}\right)\left(\frac{\Sigma x}{10}\right) + \left(\frac{\Sigma x}{10}\right)^2\right\}}$$

$$= \sqrt{\left\{\frac{\Sigma x^2}{10} - \left(\frac{\Sigma x}{10}\right)^2\right\}}.$$

It should be noted that

(i) the change of unit has not been mentioned because it is merely an artifice for removing decimals at the beginning of the calculation and putting them back at the end;

* See footnote on p. 30.

(ii) the change of origin has not been mentioned because the deviations of the observations from the mean are independent of the origin. *The general result for N observations is*:

$$\text{the standard deviation} = \sqrt{\left\{ \frac{\Sigma x^2}{N} - \left(\frac{\Sigma x}{N}\right)^2 \right\}}.$$

It can be proved by replacing 10 by N in the preceding argument.

58. Exercises

1. Working with 981 as origin and 0·1 as unit calculate the standard deviation of the six values given in §28, exercise 1.

2. By taking 40 as origin calculate the standard deviation of the ten scores of §28, exercise 2.

3. By taking 100 as origin and 0·1 as unit calculate the standard deviation of the 9 percentage relatives given in §28, exercise 3.

4. Use the data of §51, exercise 5, to calculate the standard deviation (i) for textiles, (ii) for food, and thus compare the variability of production in the two industries.

5. Use the data of §51, exercise 6, to calculate the standard deviation of (i) the beef prices, (ii) the mutton prices.

59. The standard deviation of a frequency distribution. The standard deviation of the English marks given in §46, example 3, will now be calculated. The method is an extension of that shown in §32 for calculating the mean by working with the mid-interval value of the modal class as arbitrary origin. It is shown in table 4E.

Note that Σx^2, Σx and N of §56 become $\Sigma f x^2$, $\Sigma f x$ and Σf respectively when the standard deviation of a frequency distribution is being calculated. Also, it is easier to calculate the values in the $f x^2$ column by multiplying the $f x$ value by the x value than by multiplying the f value by the square of the x value.

The student should not calculate by this method the mean and standard deviation of the Mathematics marks of the same example. The results obtained should be:

Mean　　　　　　　$= 49 \cdot 20$,
Standard deviation $= 19 \cdot 12$.

60. A comparison of the various measures of dispersion. The chief objection to the mean deviation is that it ignores plus and minus signs. This means that it does not lend itself to mathematical treatment. Nor

TABLE 4E

Deviations of
the mid-interval
mark from $44\frac{1}{2}$

Mid-interval mark	Frequency f	10 as unit x	fx		fx^2
$4\frac{1}{2}$	1	-4		$-\ 4$	16
$14\frac{1}{2}$	3	-3		$-\ 9$	27
$24\frac{1}{2}$	19	-2		$-\ 38$	76
$34\frac{1}{2}$	131	-1		-131	131
$44\frac{1}{2}$	346	0	0		0
$54\frac{1}{2}$	340	1	340		340
$64\frac{1}{2}$	134	2	268		536
$74\frac{1}{2}$	20	3	60		180
$84\frac{1}{2}$	4	4	16		64
$94\frac{1}{2}$	2	5	10		50
Total	1000	—	694	-182	1420

(i) The mean examination mark $= 44\frac{1}{2} + \dfrac{10\Sigma fx}{\Sigma f}$

$$= 44\frac{1}{2} + \frac{10 \times 512}{1000}$$

$$= 49 \cdot 62.$$

(ii) The standard deviation $= 10 \times \sqrt{\left\{ \dfrac{\Sigma fx^2}{\Sigma f} - \left(\dfrac{\Sigma fx}{\Sigma f} \right)^2 \right\}}$

$$= 10 \times \sqrt{\{1 \cdot 420 - (0 \cdot 512)^2\}}$$

$$= 10 \cdot 76.$$

is it easy to use an arbitrary origin for its calculation. The standard deviation may, at first, seem more complicated to the beginner, but it is easily calculated by working with an arbitrary origin and it is the basis of all statistical theory. Table 4F gives a summary of the advantages and limitations of the various measures of dispersion. *All are measured in the same units as the given observations.*

*61. The variance. The variance is the square of the standard deviation. If the final square root is not extracted in the calculation of the standard deviation the result obtained is then the variance. It is a measure of dispersion but it has the disadvantage of being measured in units which are the square of those of the given observations. If the heights of a group of men are given in inches, the standard deviation is in inches but the variance is in inches squared. In mathematical

* See footnote on p. 30.

theory the variance has the advantage of not involving a square root. Also certain formulae involve the square of the standard deviation.

TABLE 4F

Range	Simple idea	Based entirely on the two extremes	Unsuitable for further mathematical treatment
Interpercentile ranges	Easily found from the ogive	Ignores entirely the extremes	Unsuitable for further mathematical treatment
Semi-interquartile range or quartile deviation	Easily found from the ogive	Ignores the top and bottom quarters and is based entirely on the middle half of the observations	Unsuitable for further mathematical treatment
Mean deviation	Easily understood Working with an arbitrary origin impossible; calculation can, therefore, be tedious	Makes use of all observations but ignores signs	Unsuitable for further mathematical treatment
Standard deviation	Easily calculated by working with an arbitrary origin	Makes use of all observations; does not ignore signs	Basis of all advanced work

*62. The coefficient of variation. If two groups of observations have completely different mean values or if they are measured in completely different units their dispersion can be compared by calculating the *coefficient of variation* in each case. This is defined by

$$\text{coefficient of variation} = \frac{\text{standard deviation}}{\text{mean}},$$

or alternatively by

$$\text{coefficient of variation} = \frac{100 \times \text{standard deviation}}{\text{mean}} \%,$$

i.e. the standard deviation is expressed either as a fraction or as a percentage of the mean. The coefficient of variation is independent of the units in which the mean and standard deviation are measured. The standard deviation is an *absolute* measure, the coefficient of variation a *relative* measure of dispersion.

Example 1. The machine *A* of §46, example 1, packs cartons of sugar of mean weight 16·40 oz. and standard deviation 0·296 oz. A similar

* See footnote on p. 30.

44

machine packs 2 lb. cartons with mean weight 32·70 oz. and standard deviation 0·35 oz. Hence

the coefficient of variation of the 1 lb. machine $= 0·296/16·40$

$$= 0·018,$$

and the coefficient of variation of the 2 lb. machine $= 0·35/32·70$

$$= 0·011.$$

Thus the 2 lb. machine is less variable than the 1 lb. machine.

Example 2. The mean height of a group of men is 70 in. with standard deviation 1·3 in., while the mean weight is 160 lb. with standard deviation 7 lb. Thus

the coefficient of variation of the heights $= 1·3/70$

$$= 0·019,$$

and the coefficient of variation of the weights $= 7/160$

$$= 0·044.$$

Hence the weights are more dispersed than the heights.

63. Exercises.

1. Calculate the standard deviation of the distribution given in §26, exercise 5. Take 275½ as origin and work with 50 as unit.

2. Compare the variability of subjects A and B of §26, exercise 6, by calculating the standard deviation of each. For A, work with 65½ as origin and 10 as unit and for B, work with 35½ as origin and 10 as unit.

3. Taking 45s. as origin and 10s. as unit calculate the standard deviation of the distribution of §26, exercise 9.

4. Calculate the standard deviation of the distribution of §26, exercise 10.

*5. By calculating the coefficient of variation of the chest girths and of the heights given in §26, exercise 11, determine which of the two physical characteristics is the more dispersed.

*6. The mean *recovery time* for the aircraft of a certain fighter squadron is 18 min. with a standard deviation of 3 min., while for a certain bomber squadron the mean is 50 min. with a standard deviation of 9 min. By calculating the coefficient of variation in each case determine which squadron has the greatest variability in recovery time.

*7. The mean *length of life* of a certain type of television receiver tube is 1600 hr. with a standard deviation of 250 hr., while the mean length of life of the valves for the same receivers is 1000 hr. with standard deviation 200 hr. By calculating the coefficient of variation in each case determine which has the greater variability of life.

5

Regression

64. A bivariate distribution. The following table shows, side by side, the number of vehicles, x, on the roads of Great Britain and the total casualties, y, in road accidents for the years 1945–54. The source is the *Annual Abstract of Statistics.*

	Vehicles with licences current during the September quarter (millions)	Total casualties in road accidents (thousands)
	x	y
1945	2·6	138
1946	3·1	163
1947	3·5	166
1948	3·7	153
1949	4·1	177
1950	4·4	201
1951	4·6	216
1952	4·9	208
1953	5·3	226
1954	5·8	238

The number of vehicles is seen to have steadily increased, and the number of road accidents has also increased though not quite so regularly. The ten (x, y) pairs of values are an example of a *bivariate distribution*.

65. The scatter diagram: direct correlation. If the value x is plotted on a graph against its corresponding value y as shown in fig. 11, a *scatter diagram* is obtained.

Although the points on the scatter diagram do not fall exactly along a straight line they fall within quite a narrow belt. Small values of y correspond to small values of x, large values of y to large values of x, and x and y are said to be *directly correlated*. The exercises at the end of this chapter will provide the student with further examples of *direct correlation*.

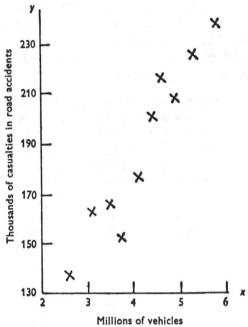

Fig. 11. Scatter diagram showing direct correlation

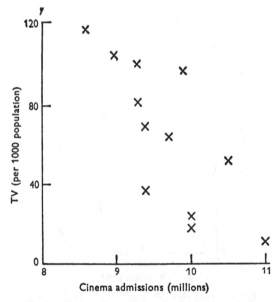

Fig. 12. Scatter diagram showing inverse correlation.

47

66. Inverse correlation. In the bivariate distribution of x and y shown in the following table, *small* values of y correspond to *large* values of x and vice versa. This is known as *inverse correlation*. In the scatter diagram (fig. 12), the points lie within a fairly well-defined belt which is downward sloping for increasing values of x.

Cinema admissions and television licences issued

Year and quarter		Towns served by Sutton Coldfield TV transmitter		Towns not normally served by any TV transmitter	
		Cinema admissions (millions) x	TV licences (per 1000 population) y	Cinema admissions (millions) z	TV licences (per 1000 population)
1950	1	11·0	12	12·1	—
	2	10·0	18	11·2	—
	3	10·0	24	11·6	—
	4	9·4	37	10·6	—
1951	1	10·5	52	12·1	—
	2	9·7	64	11·8	—
	3	9·4	69	11·5	—
	4	9·3	81	10·9	1
1952	1	9·9	98	11·6	2
	2	9·3	101	11·1	3
	3	9·0	106	11·3	3
	4	8·6	119	10·5	4
					(Northern.)

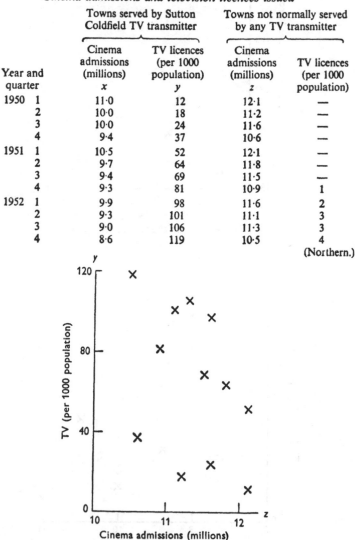

Fig. 13. Scatter diagram of a bivariate distribution in which the variables are not correlated.

67. Absence of correlation. Fig. 13 is the scatter diagram of the bi-variate distribution of y and z of the last table. One would not expect correlation in this case because it is unlikely that the *cinema admissions* in one group of towns should be related to the *number of TV licences* issued in a completely different group of towns. This is confirmed by the scatter diagram, the points of which do not lie within any well-defined belt.

68. The regression line and the regression coefficient. Fig. 14 shows the scatter diagram of fig. 11 with a straight line fitted through the middle of the ten points. It is called the *regression line*. It indicates how the *thousands of casualties, y,* vary with the *millions of vehicles, x.*

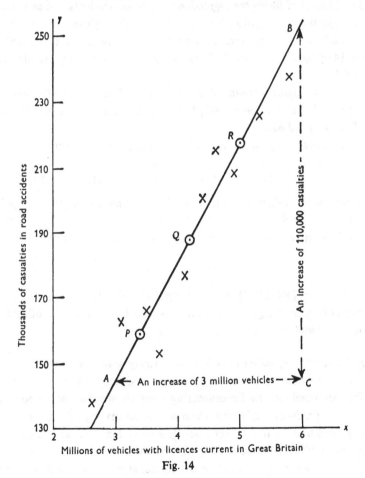

Fig. 14

49

The right-angled triangle ABC indicates that there is an increase of 110,000 casualties for an increase of 3 million vehicles. The ratio

$$\frac{CB}{AC} = \frac{110}{3} = 37 \quad \text{(correct to 2 significant figures)}$$

is known as the *slope* or *gradient* of the line and is called the *regression coefficient*. It follows that for an increase of 1 million vehicles there is an increase on the average of about 37,000 casualties. In general, the regression coefficient of y on x gives the increase in y caused by a unit increase in x. When the correlation is inverse y decreases as x increases and the regression coefficient is negative.

69. Method of fitting the regression line. Some students will be content to judge the best position for the regression line by eye. An elementary method which gives greater satisfaction, however, is illustrated in fig. 14 by the three points P, Q and R shown thus: $\odot$. They are obtained as follows:

(i) Calculate the mean values of x and y. These will be found to be 4·2 and 188·6. The point (4·2, 188·6) is Q. It is known as the *mean of the array of points.*

(ii) There are now 5 points above the mean of the array

| x | 4·4 | 4·6 | 4·9 | 5·3 | 5·8 |
| y | 201 | 216 | 208 | 226 | 238 |

whose mean values are found to be 5·0 and 217·8 and the point (5·0, 217·8) is R.

(iii) There are also 5 points below the mean of the array

| x | 2·6 | 3·1 | 3·5 | 3·7 | 4·1 |
| y | 138 | 163 | 166 | 153 | 177 |

whose mean (3·4, 159·4) is P. The regression line is drawn through the mean of the array, Q, and as nearly as possible through the other two points, P and R.

70. The two regression lines for more widely scattered points. In most scatter diagrams the points are so widely dispersed that two regression lines are possible, one for estimating y for given values of x known as *the line of regression of y on x*, the other for estimating x for given values of y known as *the line of regression of x on y*. Fig. 15 is an enlargement of fig. 12 with the two regression lines fitted. AB is the line of regression of x on y, while CD is the line of regression of y on x. AB is used to estimate the cinema admissions for a given number of TV licences per 1000 population. *CD* would be used to estimate the TV licences per 1000 population for a given number of cinema admissions. As the effect of the new form of entertainment on the old is of greater interest than the effect of the old on the new, *AB* is of greater use than *CD*.

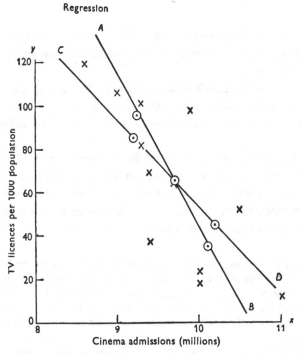

Fig. 15. Enlargement of fig. 12 (p. 47) showing the two regression lines

71. To draw the line of regression of *x* on *y*.

(i) Calculate the mean of the whole array of points (9·68, 65·08) and plot it thus: ⊙.

(ii) Now imagine that a line is drawn through (9·68, 65·08) perpendicular to the *y*-axis dividing the whole array into two separate arrays.

(iii) Calculate the means of these separate arrays (9·25, 95·7) and (10·1, 34·5) and plot them thus: ⊙.

(iv) Now draw the regression line through the mean of the whole array and as nearly as possible through the two means of the separate arrays. It shows how *x* depends on *y* and *y* is said to be the *independent variable* in this case while *x* is the *dependent variable*.

72. The line of regression of y on x. Imagine a line through the mean of the whole array (9·68, 65·08) perpendicular to the x-axis. The means of the two separate arrays thus formed are then found to be (10·18, 44·67) and (9·17, 85·5). The line of regression of y on x passes through (9·68, 65·08) and as nearly as possible through (10·18, 44·67) and (9·17, 85·5).

73. Conclusion. The above method of drawing the regression lines is a simplification of the general method which is summarized in the following statement:

The regression line should be drawn through the mean of the array in such a way that it passes reasonably near to the mean of any array of points cut off between two lines perpendicular to the axis of the independent variable.

If the two variables of a bivariate distribution are very closely correlated the points of the scatter diagram will lie within a very narrow belt and the angle between the two regression lines will be very small. For more widely dispersed points the angle between the regression lines is greater and the correlation is said to be less marked.

74. Exercises. The existence of correlation should not really be based on the evidence of less than thirty pairs of values. Dealing with such a large number of pairs is quite a long and tedious procedure. In the following exercises small numbers of pairs of values are given in order to illustrate the principles of the method without making it wearisome in its application.

1. The following table shows the examination marks of seven students in French and German. Construct a scatter diagram and draw a line of best fit:

French	12	24	30	34	47	58	68
German	32	44	47	58	73	72	88

State any conclusion that might be drawn. (London.)

2. The following table gives the marks of eight students in each of two examinations in Botany and Zoology. Construct a scatter diagram and draw a line of best fit:

Botany	39	61	49	64	42	72	52	57
Zoology	44	62	54	70	46	76	60	64

State, with a reason, any (a) general, (b) particular conclusion you could reach. (London.)

REGRESSION

3. In an experiment a vertical spring fixed at its upper end was stretched by the application of weights to its lower end and the lengths of the spring measured. The following readings were obtained:

Load (lb.)	0	1	2	3	4	5	6	7	8
Length (in.)	12	12·5	13·1	13·8	14·5	15·2	15·7	16·2	16·7

Using a scale of 1 in. to 2 lb. for the load, plot these pairs of values as points on a graph and draw the straight line which best fits them.

Assuming that the relation between the load and the length continues to hold for greater loads, find the length of the spring when the load is 12 lb.

(London.)

4. Seven students were examined in Mathematics and Physics and their marks were:

Mathematics	38	51	19	53	39	38	66
Physics	50	72	36	64	52	56	80

Construct a scatter diagram and a line of best fit. Give with reasons any (a) particular, (b) general conclusion you could reach. (London.)

5. *Index numbers of wholesale prices (X) and retail prices (Y) during* 1938 (1924 = 100)

	Jan.	Feb.	Mar.	Apr.	May	June
X	80	79	77	76	76	76
Y	84	82	81	82	81	86
	July	Aug.	Sept.	Oct.	Nov.	Dec.
X	74	72	70	71	69	70
Y	83	82	82	82	82	81

Plot a scatter diagram and draw the line of regression of Y on X.

Hence, state the average change in Y corresponding to unit positive change in X. (London.)

6. In ten areas, the percentage of dwellings overcrowded (X) and the infant mortality (Y) were as follows:

X	13	33	12	40	12	7	20	4	15	26
Y	124	151	124	156	128	78	127	104	127	144

Plot a scatter diagram, draw the line of regression of Y on X and obtain the coefficient of regression of Y on X. Explain its meaning. (London.)

7. The following table gives the percentage of sand in soil at different depths:

x (depth in in.)	0	6	12	18	24	30	36	42	48
y (% sand)	80·6	63·0	64·3	62·5	57·5	59·2	40·8	46·9	37·6

On a scatter diagram show the regression lines (labelled a and b) which you would use

(a) to predict the depth from the percentage sand;

(b) to estimate the dependence of percentage of sand on depth.

(London.)

8. A general knowledge test consisting of a hundred questions was given to fifteen boys of different ages with results as follows:

Boy	Age Years	Age Months	No. of questions correct
A	11	7	18
B	11	1	19
C	12	8	23
D	12	0	26
E	13	5	25
F	13	6	31
G	14	9	24
H	15	3	32
I	14	7	28
J	15	6	25
K	15	9	33
L	15	7	31
M	16	11	36
N	17	1	32
O	16	10	40

Plot a scatter diagram of the numbers of questions correct (y) against the age in months (x), using 1 in. to represent 10 correct questions on the y-axis and 1 in. to represent 10 months on the x-axis. Sketch in the line of regression of y on x.

State, with reasons, which boy deserves the prize for the best performance taking age into consideration. (Northern.)

9. *Percentage shrinkage in samples of cloth after washing, in directions along and across the cloth.*

Along (x)	Across (y)	Along (x)	Across (y)
12	5	7	5
4	2	12	7
10	5	18	10
10	8	14	7
11	6	14	8
10	8	8	4
6	3	11	6
6	4	17	8
6	3	21	11
13	5	12	9

Plot a scatter diagram to illustrate the data in the table, and sketch in the line of regression of x on y.

A roll of cloth is sampled by cutting a narrow test strip right across the roll. The strip proves to have a percentage shrinkage of 7. Use your regression line to obtain an estimate of the percentage shrinkage to be expected along the cloth.

It is desired to cut from the roll a piece of cloth which may be expected to shrink to 10 in. square after washing. Describe how this piece should be cut. (Northern.)

10. Five groups of locusts, each containing 120, were exposed to a lethal spray in various concentrations. The deaths resulting were as follows:

Concentration (multiple of standard)	1·2	1·4	1·6	1·8	2·0
No. of deaths	38	52	46	76	66

For each concentration, find the percentage of locusts dying. Plot this percentage against the concentration on a scatter diagram and sketch in the line of regression, taking concentration as your independent variable.

(Northern.)

11. The 1 % sample of the 1951 Census shows the ages of husband and wife to be related as shown below:

Age of wife	Age of husband					
	20–	30–	40–	50–	60–	70–
20–	12	7	—	—	—	—
30–	1	18	8	1	—	—
40–	—	2	19	6	1	—
50–	—	—	2	14	5	—
60–	—	—	—	1	8	2
70–	—	—	—	—	1	3

(1 unit = 100,000 couples)

For each range of age of husband calculate the mean age of wife. Plot your results on a regression graph and sketch in the regression line.

Estimate the mean age of wife for husband aged just 60 years.

(*Note.* The means may be calculated to the nearest year and the oldest classes may be taken as 70–80.) (Northern.)

12. The weight and height of each boy in a certain school were recorded and the pairs of measurements were found to be related as follows:

Height of boy in in.	Weight of boy in stones									
	4–	5–	6–	7–	8–	9–	10–	11–	12–	13–
51–	—	1	—	—	—	—	—	—	—	—
54–	4	12	4	—	—	—	—	—	—	—
57–	—	22	21	3	—	—	—	—	—	—
60–	—	6	28	14	1	1	—	—	—	—
63–	—	1	1	24	17	4	—	—	1	—
66–	—	—	—	4	22	18	2	3	—	—
69–	—	—	—	—	1	14	10	4	—	—
72–	—	—	—	—	—	2	2	1	—	—
75–	—	—	—	—	—	—	—	—	—	—

For each range of height calculate the mean weight in lb. Plot your results on a regression graph and sketch in the regression line.

Estimate the mean weight in lb. of boys of height 5 ft. 9 in. (Northern.)

6

Correlation by Product-moments

75. The coefficient of correlation. Correlation was first investigated graphically by Sir Francis Galton. For this reason, the scatter diagram is often called the *Galton graph*. In 1896, Karl Pearson introduced a method of assessing correlation by means of a numerical *coefficient of correlation*. This is calculated by a formula based upon a mathematical study of the regression lines. It lies between $+1$ and -1 and is positive for direct correlation and negative for inverse correlation. Values of the coefficient near to unity such as 0·85 or 0·90 indicate a high degree of correlation. Values near to zero such as 0·15 or 0·20 indicate an absence of correlation except when the coefficient has been calculated from a large number of pairs of values of the bivariate distribution. The coefficient of correlation for the *millions of vehicles*, x, and the *thousands of casualties*, y, of §64 will be calculated in this chapter and shown to be 0·96. This indicates a marked degree of correlation between the variables and the angle between the regression lines (when determined mathematically) is very small. The coefficient of correlation for the *cinema admissions*, x, and *TV licences*, y, of §66 will be shown to be $-0·75$. This indicates quite a high degree of inverse correlation. As the acute angle between the regression lines increases from 0 to 90° the numerical value of the coefficient of correlation decreases from 1 to 0. A coefficient of $+1$ indicates perfect direct correlation, a coefficient of -1 indicates perfect inverse correlation, and a coefficient of 0 indicates complete absence of correlation.

76. Calculation of the coefficient of correlation. The method of calculating the coefficient of correlation is shown in table 6A. The given values of x and y together with their totals are given in the first two columns. The next four columns are used for calculating the standard deviation, S_x, of the ten values of x, and the standard deviation, S_y, of the ten values of y. The last column contains the products of deviations, $d_x d_y$, whose mean value is called the *covariance*, S_{xy}.

77. The example of inverse correlation.. Table 6B shows the calculation of the coefficient of correlation for the *cinema admissions*, x, and

56

TABLE 6A

Calculation of coefficient of correlation (direct)

Millions of vehicles x	Thousands of casualties y	Deviations of x from the mean 4·2 d_x	Deviations of y from the mean 188·6 d_y	d_x^2	d_y^2 by tables of squares	Products of deviations $d_x d_y$
2·6	138	−1·6	−50·6	2·56	2,560·0	80·96
3·1	163	−1·1	−25·6	1·21	655·4	28·16
3·5	166	−0·7	−22·6	0·49	510·8	15·82
3·7	153	−0·5	−35·6	0·25	1,267·0	17·80
4·1	177	−0·1	−11·6	0·01	134·6	1·16
4·4	201	0·2	12·4	0·04	153·8	2·48
4·6	216	0·4	27·4	0·16	750·8	10·96
4·9	208	0·7	19·4	0·49	376·4	13·58
5·3	226	1·1	37·4	1·21	1,399·0	41·14
5·8	238	1·6	49·4	2·56	2,440·0	79·04
Total 42·0	1886	0·0	00·0	8·98	10,247·8	291·10

Covariance, S_{xy} = mean of products of deviations

$$= \Sigma d_x d_y / 10$$

$$= 29 \cdot 11.$$

Standard deviation of the 10 values of x,

$$S_x = \sqrt{\{\Sigma d_x^2 / 10\}}$$

$$= 0 \cdot 9476.$$

Standard deviation of the 10 values of y,

$$S_y = \sqrt{\{\Sigma d_y^2 / 10\}}$$

$$= 32 \cdot 01.$$

Coefficient of correlation, $r_{xy} = S_{xy} / (S_x \cdot S_y)$

$$= 29 \cdot 11 / (0 \cdot 9476 \times 32 \cdot 02)$$

$$= 0 \cdot 96 \text{ (to two places of decimals)}.$$

the *TV licences*, y, of §66. It emphasizes the need for care with + and − signs when writing down the products of deviations.

78. Terminology. By analogy with mechanics the deviations from the mean, d_x and d_y, are often called first moments about the mean, while d_x^2 and d_y^2 are called *second moments*. A convenient name for $d_x d_y$ is, therefore, *product-moment*, and r_{xy} is known as the *product-moment coefficient of correlation*.

57

TABLE 6B
Calculation of coefficient of correlation (inverse)

Cinema admissions (millions)	TV (per 1000 popula- tion)	Deviations of x from the mean 9·68	Deviations of y from the mean 65·08	By four-figure tables		
x	y	d_x	d_y	d_x^2	d_y^2	$d_x d_y$
11·0	12	1·32	−53·08	1·7420	2,818·0	− 70·07
10·0	18	0·32	−47·08	0·1024	2,217·0	− 15·07
10·0	24	0·32	−41·08	0·1024	1,688·0	− 13·15
9·4	37	−0·28	−28·08	0·0784	788·5	+ 7·86
10·5	52	0·82	−13·08	0·6724	171·1	− 10·73
9·7	64	0·02	− 1·08	0·0004	1·2	− 0·02
9·4	69	−0·28	3·92	0·0784	15·4	− 1·10
9·3	81	−0·38	15·92	0·1444	253·4	− 6·05
9·9	98	0·22	32·92	0·0484	1,083·0	+ 7·24
9·3	101	−0·38	35·92	0·1444	1,290·0	− 13·65
9·0	106	−0·68	40·92	0·4624	1,675·0	− 27·83
8·6	119	−1·08	53·92	1·1660	2,907·0	− 58·23
*Total 116·1	781	−0·06	0·04	4·7420	14,907·6	−200·80

Covariance, $\qquad\qquad S_{xy} = \Sigma d_x d_y / 12$

$$= -16·73.$$

Standard deviation, $\qquad S_x = \sqrt{\{\Sigma d_x^2 / 12\}}$

$$= 0·6286.$$

Standard deviation, $\qquad S_y = \sqrt{\{\Sigma d_y^2 / 12\}}$

$$= 35·25.$$

Coefficient of correlation, $\qquad r_{xy} = S_{xy} / (S_x S_y)$

$$= -16·76 / (0·6319 \times 35·24)$$

$$= -0·76 \text{ (to two places of decimals)}.$$

79. Table of minimum numerical values of r_{xy}. To establish that correlation exists between two variables, r_{xy} should be calculated from as large a number of pairs of observations as possible. If the numerical values of r_{xy} (i.e. $|r_{xy}|$) are equal to the values shown in the right-hand column of table 6C, the mathematical probability of correlation is 19/20 (i.e. there is a 19 to 1 chance of association between the two variables). Larger values of $|r_{xy}|$ mean that the probability of correlation is even greater.

* The totals of the d_x and d_y columns would, of course, be zero if the means were exact.

PRODUCT MOMENTS

TABLE 6C

No. of pairs of values of x and y from which r_{xy} is calculated	Minimum value of $|r_{xy}|$ for correlation to be probable
6	0·82
7	0·76
8	0·71
9	0·67
10	0·64
11	0·61
12	0·58
13	0·56
14	0·54
15	0·52
16	0·50
18	0·47
20	0·45
40	0·31
80	0·22
100	0·20

80. Exercises.

1–10. Calculate the product-moment coefficients of correlation of the bivariate distributions given in exercises 1–10, §74.

11. From the following data calculate the correlation coefficient between the wheat yield and the area planted in England and Wales for the years 1929–38.

Year	Wheat yield (millions of tons)	Area planted (millions of acres)
1929	1·3	1·4
1930	1·1	1·4
1931	1·0	1·2
1932	1·1	1·3
1933	1·6	1·7
1934	1·8	1·8
1935	1·6	1·8
1936	1·4	1·7
1937	1·4	1·7
1938	1·9	1·8

(Northern)

12. Compute the coefficient of correlation between the following series of index numbers for Great Britain:

Year	Quarters	Board of Trade	
		Wholesale prices	Physical volume of production
1931	1	64	85
	2	63	81
	3	60	81
	4	64	91
1932	1	63	91
	2	61	83
	3	60	78
	4	61	87
1933	1	60	88
	2	60	89

What meaning do you attach to your result? (London.)

13. Calculate the coefficient of correlation for the following series:

United Kingdom
index numbers (1938 = 100)

1947	Import prices X	Export prices Y
Jan.	223	211
Feb.	226	215
Mar.	229	220
Apr.	234	221
May	241	225
June	245	227
July	252	230
Aug.	254	234
Sept.	254	236
Oct.	256	237

What meaning do you attach to your result? (London.)

PRODUCT MOMENTS

14. *Index Numbers of the cost of living (X) and wages rates (Y),*
United Kingdom

	X	Y
	(Ministry of Labour)	(Bowley)
	(av. 1924= 100)	(Dec. 1924= 100)
1937	89	101
1938	89	104
1939	91	106
1940	101	117
1941	114	127
1942	114	136
1943	114	144
1944	115	153
1945	116	161
1946	116	175

Calculate the coefficient of correlation for these series.
What light does your result throw on the relationship between the two
series?   (London).

15. *Interim index of production (X) and index of volume of exports (Y).*
Monthly averages. United Kingdom (Monthly Digest of Statistics)

Year	Quarter	X (av. 1946= 100)	Y (1938= 100)
1946	1	94	84
	2	98	98
	3	99	104
	4	109	112
1947	1	96	101
	2	109	102
	3	109	114
	4	119	118
1948	1	120	126
	2	122	134

Calculate the coefficient of correlation for these series.
What meaning do you attach to your result?   (London.)

7

Correlation by Ranks

81. The coefficient of rank correlation. The following table shows the data of §64 with *ranks* attached to the x and y values:

No. of vehicles (millions) x	Rank R_x of the x values	No. of casualties (thousands) y	Rank R_y of the y values
2·6	10	138	10
3·1	9	163	8
3·5	8	166	7
3·7	7	153	9
4·1	6	177	6
4·4	5	201	5
4·6	4	216	3
4·9	3	208	4
5·3	2	226	2
5·8	1	238	1

The number 2·6, for example, is *tenth* in *order* or *rank* of the x values, 4·1 is *sixth* and 5·8 is *first*. If the product-moment coefficient of correlation is calculated from the ranks R_x and R_y instead of the original values of x and y far less arithmetic is involved and quite a good approximation to r_{xy} is obtained which is called the *coefficient of rank correlation*, R. This idea was originally introduced by C. Spearman in 1906. By applying the method shown in table 6A to the R_x and R_y values the student will obtain $R=0·95$.

82. The formula for the coefficient of rank correlation. The ranks are the numbers 1, 2, 3, 4, ..., n (n in the example under consideration being 10), and the algebraic formulae for the sum of the first n natural numbers and for the sum of their squares can be used to prove that R may be calculated as shown in table 7A by the formula

$$R = 1 - \frac{6\Sigma D^2}{n(n^2-1)},$$

where D is the *rank difference*.

Calculation of the coefficient of rank correlation (direct)

Rank of no. of vehicles R_x	Rank of no. of casualties R_y	Rank difference $R_x - R_y$ D	D^2
10	10	0	0
9	8	1	1
8	7	1	1
7	9	−2	4
6	6	0	0
5	5	0	0
4	3	1	1
3	4	−1	1
2	2	0	0
1	1	0	0
	Total	0	8

Since $n = 10$ and $\Sigma D^2 = 8$,

$$R = 1 - \{6 \times 8/10(10^2 - 1)\}$$
$$= 0.95.$$

83. Method of ranking equal values of a variate. If the *cinema admissions (millions)* of §66 are arranged in descending order of magnitude they appear as follows:

11·0, 10·5, 10·0, 10·0, 9·9, 9·7, 9·4, 9·4, 9·3, 9·3, 9·0, 8·6.

The rank of 11·0 is, therefore, 1 and that of 10·5 is 2. The rank of the *two* values 10·0 is not taken as 3 or 4 but as 3·5, the mean of the ranks 3 and 4. Similarly, the ranks of the two values 9·4 is taken as 7·5 and that of the two values 9·3 as 9·5. The twelve ranks are thus:

1, 2, 3·5, 3·5, 5, 6, 7·5, 7·5, 9·5, 9·5, 11, 12

and their sum is 78, the same as the sum of the ranks

1, 2, 3, 4, 5, 6, 7, 8, 9, 10, 11, 12

when the values of the variate are all different.

Table 7B shows the method of calculating the coefficient or rank correlation for the *cinema admissions, x,* and *TV licences y* of §66. It is a good example of inverse correlation. The value obtained, $R = -0.84$, is considerably higher (numerically) than the value $r_{xy} = -0.75$ calculated in table 6B. It must be clearly understood that table 6C is for minimum values of $|r_{xy}|$. No similar table exists for minimum values of $|R|$. Indeed, R is nothing more than a quickly calculated approximation to r_{xy}.

TABLE 7B

Calculation of the coefficient of rank correlation (inverse)

Cinema admissions		TV licences per 1000 population		Rank difference $R_x - R_y$	
(millions) x	Rank R_x	y	Rank R_y	D	D^2
11·0	1	12	12	−11	121
10·0	$3\frac{1}{2}$	18	11	$-7\frac{1}{2}$	$56\frac{1}{4}$
10·0	$3\frac{1}{2}$	24	10	$-6\frac{1}{2}$	$42\frac{1}{4}$
9·4	$7\frac{1}{2}$	37	9	$-1\frac{1}{2}$	$2\frac{1}{4}$
10·5	2	52	8	− 6	36
9·7	6	64	7	− 1	1
9·4	$7\frac{1}{2}$	69	6	$1\frac{1}{2}$	$2\frac{1}{4}$
9·3	$9\frac{1}{2}$	81	5	$4\frac{1}{2}$	$20\frac{1}{4}$
9·9	5	98	4	1	1
9·3	$9\frac{1}{2}$	101	3	$6\frac{1}{2}$	$42\frac{1}{4}$
9·0	11	106	2	9	81
8·6	12	119	1	11	121
			Total	0	$526\frac{1}{2}$

Since $n = 12$ and $\Sigma D^2 = 526\frac{1}{2}$

$$R = 1 - \{6 \times 526\tfrac{1}{2}/12(12^2 - 1)\}$$
$$= -0·84.$$

84. Exercises.

1–15. Calculate the coefficients or rank correlation for the bivariate distributions of exercises 1–15, §80. Note that they are approximately equal to their respective product-moment coefficients of correlation.

16. Ten shades of the colour green when arranged in their true order from light to dark are numbered 1–10 respectively. An observer, when asked to arrange the shades from light to dark, produces the following rank

3, 1, 5, 2, 6, 4, 10, 9, 7, 8.

What is the value of Spearman's coefficient of rank correlation in this case?
(London.)

RANK CORRELATION

17.   *The County Cricket Championship Table.*

Final position

	1956	1955
Surrey	1	1
Lancashire	2	9½
Gloucester	3	12
Northants	4	7
Middlesex	5	5
Hampshire	6	3
Yorkshire	7	2
Notts	8	11
Sussex	9½	4
Worcester	9½	15
Essex	11	14
Derbyshire	12	8
Glamorgan	13	16
Warwick	14	9½
Somerset	15	17
Kent	16	13
Leicester	17	6

Calculate the coefficient of correlation between the ranks of the County Cricket Clubs shown above.

18. Explain briefly what is meant by correlation.

A class of 15 students was examined in Mathematics and Physics. The following table gives the orders of merit of the students (arranged in alphabetical order) in both subjects. Calculate the coefficient of ranked correlation:

Mathematics	11	5	9	14	13	2	6	10	1	4	7	3	12	8	15
Physics	11	8	7	9	12	1	4	10	3	6	14	2	13	5	15

(London.)

19. There are ten finalists in a competition for which there are two judges X and Y. The following table gives the order in which X and Y place the competitors. Calculate the rank correlation coefficient and state any conclusions that you deduce from it:

Competitors	A	B	C	D	E	F	G	H	K	L
Ranking of X	6	10	2	7	1	4	5	9	8	3
Ranking of Y	2	7	8	10	6	1	9	5	3	4

(London.)

20. In a drama competition ten plays were ranked by two adjudicators as follows:

Play	A	B	C	D	E	F	G	H	J	K
Rank given by X	5	2	6	8	1	7	4	9	3	10
Rank given by Y	1	7	6	10	4	5	3	8	2	9

Calculate the coefficient of ranked correlation. Is there any reason for saying that there is a significant agreement between the two adjudicators?

(London.)

65

8

The Analysis of a Time-series

85. The graph of a time-series, or historigram. A time-series is a series of values assumed by a variable at different points of time. The *cinema admissions, x,* of §66 is a good example of such a series. Noticeable features of the cinema admission figures are:

(i) For the first quarter of each year they are high compared with the other three quarters. This is an example of *seasonal variation* in the series.

(ii) During the three years they decrease and we say that the *general trend* of the series is downwards.

The graph of the time-series shown in fig. 16 indicates the seasonal variation of the *peaks* P_2, P_3, and the *troughs* T_1, T_2.* It gives the *history* of the cinema admissions during the period 1950–2 and is called the *historigram*. The *four-quarterly moving averages* graph included in the same diagram indicates the general trend.

86. The four-quarterly moving averages. If

$m_1 =$ the mean of the cinema admissions for quarters 1, 2, 3 and 4,

$m_2 =$ the mean of the cinema admissions for quarters 2, 3, 4 and 5,

$m_3 =$ the mean of the cinema admissions for quarters 3, 4, 5 and 6, etc.,

then m_1, m_2, m_3, etc., are called the four-quarterly moving averages of the cinema admissions. The smoother graph obtained by plotting the moving averages gives a better idea of the general trend than the graph of the unmodified series.

87. A convenient method of calculating the moving averages. In the example under consideration

$$m_1 = \tfrac{1}{4}(11 \cdot 0 + 10 \cdot 0 + 10 \cdot 0 + 9 \cdot 4)$$
$$= 10 \cdot 1.$$

On the graph it is plotted in the middle of the $1-4$ quarter range.

---

* There is no justification for saying that P_1 is a peak and T_3 a trough unless the preceding and following figures are known.

Instead of calculating m_2 by $m_2 = \frac{1}{4}(10\cdot0 + 10\cdot0 + 9\cdot4 + 10\cdot5)$ it is more convenient to use

$$m_2 = m_1 + \frac{1}{4}(10\cdot5 - 11\cdot0)$$
$$= 10\cdot1 - 0\cdot125$$
$$= 9\cdot975.$$

Similarly, $m_3 = m_2 + \frac{1}{4}(9\cdot7 - 10\cdot0)$, and so on for m_4, m_5, etc. Table 8A shows a convenient method of setting out the whole calculation. The *table of differences* is obtained by subtracting each x value from the x value directly below it. The differences are then divided by 4 and

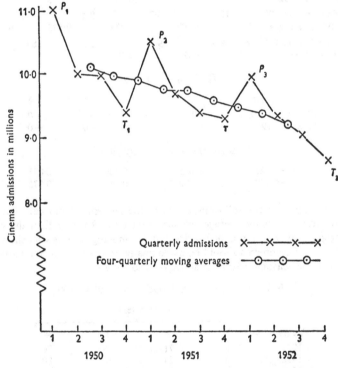

Fig. 16. Historigram of the unmodified series and the four-quarterly moving averages. (The zigzag in the ordinate scale indicates that it does not begin at zero.)

entered in the third section of the table. The values in this third section are then *added cumulatively* to the mean of the first four quarters, 10·100, to obtain the series of four-quarterly moving averages:

10·100, 9·975, 9·900, 9·750, 9·725, 9·575, 9·475, 9·375, 9·200.

These four-quarterly moving averages are finally written in the first section of the table, as shown, in the middle of their respective four-quarterly ranges. Note particularly that the moving averages are entered in the first section of the table after the second and third sections have been completed.

TABLE 8A
Calculation of four-quarterly moving averages

Cinema admissions, x

Quarter ...	1	2	3	4
Admissions in 1950, x	11·0	10·0	10·0	9·4
Moving averages		10·100	9·975	9·900
Admissions in 1951, x	10·5	9·7	9·4	9·3
Moving averages	9·750	9·725	9·575	9·475
Admissions in 1952, x	9·9	9·3	9·0	8·6
Moving averages	9·375	9·200		

Table of differences

Quarter ...	1	2	3	4
1951, x −1950, x	−0·5	−0·3	−0·6	−0·1
1952, x −1951, x	−0·6	−0·4	−0·4	−0·7

Differences divided by four

−0·125	−0·075	−0·150	−0·025
−0·150	−0·100	−0·100	−0·175

88. Exercises. In the following exercises the student should calculate the four-quarterly moving averages as shown in table 8A and then draw the historigram of the unmodified series and the moving averages as shown in fig. 16. Brief comments should be made finally on the general trend and the extent of the seasonal variation above or below the trend.

1.   *Index numbers of retail food prices* (1924 = 100)

Quarters

Years	1	2	3	4
1935	72	71	76	76
1936	75	74	77	80
1937	80	81	83	85

(London.)

2.   *Quarterly index numbers of coal production* (1924 = 100)

Quarters

Years	1	2	3	4
1936	93·3	81·7	81·5	89·1
1937	93·8	92·3	86·5	93·7
1938	97·6	82·2	79·0	89·3

(London.)

3. *Percentages of insured workpeople unemployed,*
United Kingdom (quarterly averages)

	Quarters			
Years	1	2	3	4
1925	11·2	10·9	12·0	11·0
1926	10·4	14·3	14·0	13·5
1927	10·9	8·8	9·3	9·8
1928	10·4	9·8	11·5	11·7

(London.)

Note the *abnormal* rise in 1926 due to the *General Strike* which is not a seasonal variation.

4. *Production of passenger cars, U.S.A. (tens of thousands)*
(quarterly medians)

	Quarters			
Years	1	2	3	4
1927	26	36	24	11
1928	29	36	36	22
1929	40	52	43	17

(London.)

5. *Personal expenditure on fuel and light, United Kingdom, £ million*

	Quarters			
Years	1	2	3	4
1946	84	64	61	82
1947	92	70	63	85
1948	100	81	72	96

(London.)

Source: *Monthly Digest of Statistics*, April 1949.

6. *Number of licences current for road vehicles in Great Britain*
at the end of the month stated (tens of thousands)

	1947	1948	1949	1950	1951	1952
Feb.	295	261	350	383	393	416
May	329	315	386	419	436	462
Aug.	344	365	402	431	452	477
Nov.	325	368	405	432	449	475

(London.)

89. The twelve-monthly moving average. Table 8B shows the monthly production figures of the manufacturing industries of the United Kingdom (Source: *Monthly Digest of Statistics*). Seasonal variations in this series recur at twelve-monthly intervals. They are eliminated by the twelve-monthly moving averages shown in the same table. The calculation shown in table 8B is an obvious extension of that of table 8 A to groups of twelve instead of four. The historigram (fig. 17), shows that the seasonal variation takes the form of a definite drop in production each year during the summer holiday period. The moving average graph shows a steady increase in the general trend of about 10 points per annum (i.e. 10 % of the production of 1946).

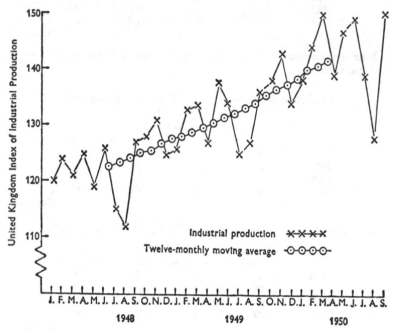

Fig. 17. Historigram of the monthly index of industrial production and the 12-monthly moving averages.

90. The seven-day moving average. In the daily attendances at an exhibition given in table 8 C a rise and fall due to the day of the week is noticeable. On Sunday the attendance is low, on Monday it is high and on Tuesday it is low again. During the rest of the week the attendances steadily increase until the maximum is reached on Saturday. These same fluctuations recur in the second week and they are eliminated by the seven-day moving average which is calculated in table 8 C and illustrated graphically in fig. 18. The general trend shows a very slight increase in attendances during the two weeks.

91. The ten-yearly moving average. Table 8 D shows the total number of births in the United Kingdom each year for a period of 22 years and

TABLE 8B

Calculation of twelve-monthly moving averages

Index of industrial production, 1946= 100 (all manufacturing industries)

	Jan.	Feb.	Mar.	Apr.	May	June	July	Aug.	Sept.	Oct.	Nov.	Dec.
1948 indices	120	124	121	125	119	126	115	112	127	128	131	125
Moving averages						122·75	123·25	124·00	125·08	125·25	126·83	127·50
1949 indices	126	133	134	127	138	134	120	122	136	138	143	134
Moving averages	127·92	128·75	129·50	130·33	131·33	132·08	133·08	134·00	135·33	136·33	137·08	138·33
1950 indices	138	144	150	139	147	149	139	128	150			
Moving averages	139·92	140·42	141·58									

Table of differences

	Jan.	Feb.	Mar.	Apr.	May	June	July	Aug.	Sept.	Oct.	Nov.	Dec.
1949 − 1948	6	9	13	2	19	8	5	10	9	10	12	9
1950 − 1949	12	11	16	12	9	15	19	6	14			

Differences divided by twelve (to two decimals)

	Jan.	Feb.	Mar.	Apr.	May	June	July	Aug.	Sept.	Oct.	Nov.	Dec.
1949 − 1948	0·50	0·75	1·08	0·17	1·58	0·67	0·42	0·83	0·75	0·83	1·00	0·75
1950 − 1949	1·00	0·92	1·33	1·00	0·75	1·25	1·58	0·50	1·17			

TABLE 8C

Calculation of seven-day moving averages

Daily attendances, in thousands, at an exhibition

	Sun.	Mon.	Tues.	Wed.	Thurs.	Fri.	Sat.
1st week	35	70	36	59	62	68	71
Moving averages				57·29	57·86	58·14	58·43
2nd week	39	72	38	56	63	81	75
Moving averages	58·00	58·14	58·57	59·14			

Table of differences

2nd week – 1st week	4	2	2	−3	1	3	4

Differences divided by seven (to two decimals)

0·57	0·29	0·29	−0·43	0·14	0·43	0·57

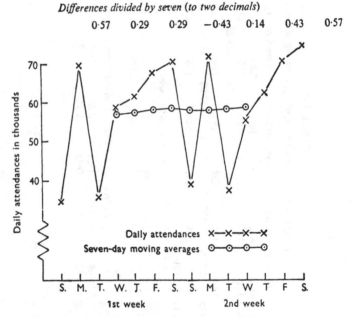

Fig. 18. Daily attendances and the seven-day moving averages.

fig. 19 is the historigram. Notice that there is a rise and fall in the period 1933–41, a second rise and fall in the period 1941–5 and a third rise and fall in the period 1946–52. These are examples of *cyclical variations*. An upward trend over a period of twenty or thirty years is usually attained, not by a steady increase each year, but by cycles which vary in period from five to ten years. The ten-yearly moving averages, which are calculated in table 8D and illustrated graphically in fig. 19,

TABLE 8D

Calculation of ten-yearly moving averages

Total number of births, in thousands, in the United Kingdom between the years 1933 and 1954

	1933	1934	1935	1936	1937	1938	1939	1940	1941	1942
Births	692	712	711	720	724	736	727	702	696	772
Moving averages					719·2	731·1	747·7	756·2	779·7	809·8

	1943	1944	1945	1946	1947	1948	1949	1950	1951	1952
Births	811	878	796	955	1025	905	855	818	797	793
Moving averages	826·7	839·5	851·1	861·2	863·3	862·6	854·3			

	1953	1954
Births	804	795
Moving averages		

Table of differences

	119	166	85	235	301	169	128	116	101	21
	−7	−83								

Differences divided by ten

	11·9	16·6	8·5	23·5	30·1	16·9	12·8	11·6	10·1	2·1
	−0·7	−8·3								

remove the cyclical variations and reveal an upward trend in the birth-rate until 1947 and the beginning of a decline after 1947. The ease with which the differences can be divided by ten is the reason for working with ten-yearly moving averages rather than, say, seven-yearly moving averages. Five-yearly moving averages are often used. They offer almost the same convenience in division as ten-yearly moving averages but do not always completely eliminate the cyclical variations.

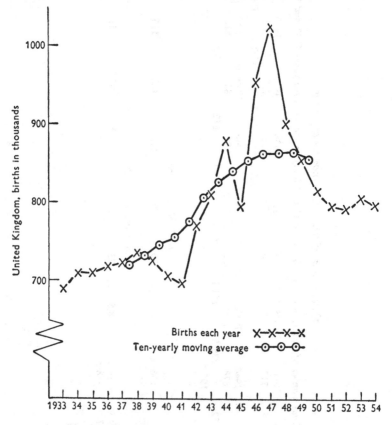

Fig. 19. The rising birth-rate of the United Kingdom.

92. Summary of the analysis of a time-series. In the analysis of a time-series it is necessary to distinguish between the following types of variation:

(i) *The general trend* over several decades. It is also called the *secular variation.*

(ii) *Cyclical variations* which are fluctuations in the general trend. Their period vary usually from five to ten years. They are eliminated by the ten-yearly moving average.

(iii) *Seasonal variations* which recur annually due to the seasons of the year. They are eliminated by the four-quarterly or twelve-monthly moving average.

(iv) *Special variations* due to abnormal circumstances. The high unemployment figures in the second and third quarters of 1926 (see §88, exercise 3) due to the *General Strike* and the high birth-rate in 1947 after *World War II* (see table 8 D) are examples of special variations.

An example has also been given of variations due to the day of the week. These were removed by taking seven-day moving averages. It is equally possible that, in a time-series consisting of weekly figures, variations may occur due to the week of the month. Such variations could be conveniently removed by four-weekly or five-weekly moving averages.

93. Exercises.

1. Explain what is meant by 'moving average'.

The following table gives the numbers (in 1000's) of passenger cars and chassis produced for export in the months from April 1948 to June 1950. Construct a table showing the twelve-monthly moving averages and draw a graph to show the trend:

	J	F	M	A	M	J	J	A	S	O	N	D
1948	.	.	.	20	20	25	18	16	25	18	17	19
1949	18	19	26	18	21	26	16	20	26	24	28	28
1950	32	33	41	29	34	40	.	.	.	.	.	.

(London.)

2. The table below shows the number of permanent houses completed in England and Wales each month from January 1952 to August 1953, the numbers being in thousands:

	J	F	M	A	M	J	J	A	S	O	N	D
1952	13·5	14·6	19·0	15·4	17·3	17·0	18·2	15·6	19·7	20·5	19·5	19·7
1953	17·8	17·5	25·0	20·2	22·8	22·5	24·9	24·7	.	.	.	.

Plot these figures on a graph. Calculate the 12-month moving average and plot its graph with your previous one. (London.)

3. A Ballet Company gave a six weeks' season at a large hall capable of seating 4000 people and the attendances in hundreds, at the evening performances, are recorded in the following table:

Attendances, in hundreds, to nearest hundred

	Mon.	Tues.	Wed.	Thurs.	Fri.	Sat.
First week	12	13	20	19	17	24
Second week	22	25	31	31	26	34
Third week	30	31	40	38	36	40
Fourth week	40	40	40	40	40	40
Fifth week	38	39	40	40	40	40
Sixth week	32	33	34	32	30	36

Plot a graph of the above time-series and include on the same diagram the graph of the *six-day* moving averages.

Comment on the weekly *cycle* of attendances and state, with reasons, if you think an extension of the season to eight weeks would have been justified.

(Northern.)

4. Draw roughly (not on graph paper) a diagram to represent a series of twelve annual figures for the price of a commodity, satisfying the following conditions: the general tendency during the period was for the price to rise, but for four separate years the price was lower than for the preceding year. Point out the features of your graph that satisfy the conditions. (Northern.)

5. Give an account of the analysis of a time-series indicating why the ordinary statistical averages are not suitable in this case. Explain the use of the method of moving averages to isolate the general trend of such a series.

(London.)

6. What do you understand by cyclical variations in trade and business?

As a manager of a sales organization, describe briefly how you would estimate next year's sales. You may assume, for illustrative purposes, that your sales organization deals with any well-known commodity or type of commodity. (London.)

7. Plot a graph showing the changes with time in the following index numbers of real wages and comment on the nature of the fluctuations:

Year	Index no.	Year	Index no.
1880	75	1890	95
1881	78	1891	93
1882	78	1892	90
1883	81	1893	90
1884	78	1894	92
1885	87	1895	95
1886	79	1896	99
1887	84	1897	98
1888	88	1898	98
1889	91	1899	102

Calculate the ten-yearly moving averages and plot them on the same diagram. Comment on the general trend of real wages during this period.

(Northern.)

8. From the following data calculate the five-yearly moving averages of the prices of beef and mutton for the years 1925–38 and plot their two graphs in one diagram.

Comment on any correlation in the trend of prices of the two commodities.

Price in pence per 8 lb.

Year	Beef	Mutton
1925	80	106
1926	74	89
1927	70	86
1928	74	92
1929	71	89
1930	73	92
1931	67	79
1932	65	63
1933	61	69
1934	58	74
1935	54	75
1936	54	73
1937	61	78
1938	62	62

(Northern.)

9

Weighted Averages

94. Percentage relatives. If the prices of beef and mutton given in §93, exercise 8, are expressed as percentages of the price in 1925 they are called *percentage relatives with* 1925 *as the base year*. Table 9 A shows the complete time-series transformed into percentage relatives (to the nearest whole number). The rapid drop in the price of mutton compared with that of beef is much more apparent from the percentage relatives than from the actual prices. The percentage relatives of prices are often called *price relatives*.

TABLE 9A

Percentage relatives with 1925 as base

Year	Beef	Mutton
1925	100	100
1926	93	84
1927	88	81
1928	93	87
1929	89	84
1930	91	87
1931	84	75
1932	81	59
1933	76	65
1934	73	70
1935	68	71
1936	68	69
1937	76	74
1938	78	58

95. Weighted average. If we wish to study the variation in the price of meat in general we might take the average of 93 and 84 (i.e. 88½) as an indication of the price in 1926. Research by H.M. Cost of Living Advisory Committee shows, however, that for every 24 oz. of beef consumed only 14 oz. of mutton are consumed. This means that an alteration in the price of mutton will not have such a great effect on the amount of money spent at the butcher's shop as an alteration in the price of beef. In fact it seems obvious that to get a true indication of the change in the price of meat we should multiply the price of beef by the weight of beef consumed and the price of mutton by the weight of mutton consumed and divide the sum by the total weight. Thus for 1926 we should get

$$\frac{(93 \times 24 + 84 \times 14)}{(24 + 14)} = \frac{3408}{38}$$

$$= 89\cdot7.$$

The result of 89·7 thus obtained is called the *weighted average* of the beef and mutton price relatives. It is a *composite* price relative for beef and mutton.

A complete indication of the change in price of meat, however, should also take into account the change in price of pork, sausages, pies, etc., and table 9B reproduces the weights assigned by the Cost of Living Advisory Committee based on the consumption in 1953–4. Thus if b, m, p and s are the respective price relatives of beef, mutton, pork and sausages, etc., the weights given in table 9B mean that the weighted average of b, m, p and s is

$$\frac{(24b + 14m + 6p + 25s)}{(24 + 14 + 6 + 25)}.$$

TABLE 9B

	Weight
Beef	24
Mutton and lamb	14
Pork	6
Sausages, pies, canned meat, offal, poultry, etc.	25

96. The arithmetical calculation of a weighted average. The calculation of a weighted average can be written out in exactly the same way as the calculation of the mean of a frequency distribution shown in §30, the *percentage relatives* and the *weights* taking the place respectively of the *numbers of tomatoes per truss* and the *frequencies*. If the price relatives of beef, mutton, pork and sausages, etc., in 1926 are 93, 84, 88 and 82 respectively it would be convenient to work with 84 as origin and to tabulate the calculation as shown in table 9C.

97. Index numbers. The weighted average (or composite price relative) 86·8 obtained in table 9C is an example of an *index number*. It indicates that the price of meat in 1926 is 86·8 % of the price in the base year 1925. Alternatively, we might say that the price in 1926 was 13·2 % below the price in 1925. The number 86·8 might be called the *price index* of meat. Further examples of index numbers will be given in this chapter. Indeed, some were given in chapter 8. The retail food price index and the coal production index (with 1924 as base year) were used in §88, exercises 1 and 2, and the index of industrial production (with 1946 as base year) was used in §89, table 8 B.

TABLE 9C

Calculation of a weighted average

	Price relative	Weight w	Price relative with 84 as origin x	wx
Beef	93	24	9	216
Mutton and lamb	84	14	0	0
Pork	88	6	4	24
Sausages, pies, canned meat, offal, poultry, etc.	82	25	−2	−50
	Total	69	Total	190

$$\text{Weighted average} = 84 + \Sigma wx / \Sigma w$$
$$= 84 + 190/69$$
$$= 86\cdot8.$$

98. The choice of base. It is always difficult to decide which year will prove the most satisfactory base. For the *cost-of-living index* (an index of the greatest interest to all of us) conditions prior to the outbreak of World War I formed the standard of reference for many years, July 1914 being taken as the base. By the end of World War II this cost-of-living index had risen to 203, that is to say, prices in 1947 were more than twice what they were in 1914. Moreover, it was realized that the method of calculating the cost-of-living index was completely out of date (see §99). So in June 1947 it was replaced by the *interim index of retail prices*, June 1947 being taken as base. The word 'interim' was used because the new index was a temporary measure to be used until a permanent measure, based on research carried out by the Cost of Living Advisory Committee could be established. This permanent *index of retail prices* was finally established in 1956 and January 1956 was taken as base. By October 1956 this new index had risen to 102·7. If June 1947 had continued as base it would have been 158, and if July 1914 had continued as base it would have been well over 300. Thus, the cost of living in October 1956 was 2·7 % above January 1956, 58 % above June 1947 and over 200 % above July 1914. This example serves to show the way in which the base for an index number was chosen and altered later when it was found necessary to do so.

99. The index of retail prices. The calculation of the index of retail prices is shown in table 9 D. The price relative 101·8 for food indicates that the price of food on 16 October 1956 was 1·8 % above the price on 17 January 1956. The number 101·8 is calculated by extending a table such as table 9 C to include not only meats but all items of food such as bread, flour, bacon, fish, butter, cheese, milk and eggs. The price relative 104·5 for housing is calculated by considering the rises in rent and rates, charges for repairs, etc.; the 103·5 for miscellaneous goods by considering the price of books, newspapers, stationery, medicines, toilet requisites, soap, fancy goods and toys; the 103·1 for transport and vehicles by considering railway fares, bus fares, cost of cycles, etc.; and the 105·6 for services by considering postages, entertainments, laundry, hairdressing, etc.

TABLE 9D

Calculation of the index of retail prices for 16 *October* 1956

(17 *January* 1956 = 100)

Group	Price relative 16 Oct. 1956	Weight w	Price relative with 100 as origin x	wx
Food	101·8	350	1·8	630
Alcoholic drink	102·6	71	2·6	184·6
Tobacco	105·3	80	5·3	424
Housing	104·5	87	4·5	391·5
Fuel and light	102·4	55	2·4	132
Durable household goods	101·3	66	1·3	85·8
Clothing and footwear	101·0	106	1·0	106
Transport and vehicles	103·1	68	3·1	210·8
Miscellaneous goods	103·5	59	3·5	206·5
Services	105·6	58	5·6	324·8
	Total	1000	Total	2696·0

Index of retail prices $= 100 + \Sigma wx / \Sigma w$

$$= 100 + 2696/1000$$

$$= 102·7 \ (1 \ \text{decimal}).$$

The weights of the second column need some explanation because they are not really weights but estimates of the proportional importance of each group. They might be interpreted by saying that out of every thousand shillings 350 are spend on food, 71 on alcoholic drink, etc. Alternatively, it can be said that out of every £5 the amounts spent respectively on food, drink, tobacco, housing are 35s., 7s. 1d., 8s., 8s. 8d. The weights result from research into the expenditure of a *random sample* (see §109) of 11,638 households of the United Kingdom. They are called 'weights' because the calculation of the index of retail prices as a weighted average of the price relatives is an extension of the simple idea developed in §95 in which the relative importance of the changes in price of beef, mutton and pork is dependent on the weights of each type of meat consumed.

It has already been mentioned that the cost-of-living index which had been in operation since July 1914 was completely out-of-date in 1947. The reason for this was that people were spending their money differently in 1947. New kinds of food, new entertainments, better forms of transport and completely different fashions and materials for clothing had transformed their way of life. Even since 1947 the amounts of money spent on the different groups in the left-hand column of table 9 D have altered and table 9 E shows how the weights were adjusted by the Cost of Living Advisory Committee in the interim period between 1947 and 1956. The amounts spent on clothing, transport, services and miscellaneous goods have noticeably increased, while the amounts spent on drink and tobacco have decreased.

TABLE 9E

The adjustment, between 1947 and 1956, of the weights used in the calculation of the index of retail prices

	Weights used from 17 June 1947 to 15 Jan. 1952	Weights used from 15 Jan. 1952 to 17 Jan. 1956 based on estimates of consumption in 1950	Weights used from 17 Jan. 1956 based on ascertained consumption in 1953–4
Food	348	399	350
Alcoholic drink	} 217	{ 78	71
Tobacco		{ 90	80
Housing	88	72	87
Fuel and light	65	66	55
Durable household goods	71	62	66
Clothing and footwear	97	98	106
Transport and vehicles	} 79	91	{ 68
Services			{ 58
Miscellaneous goods	35	44	59

100. The crude and standardized death-rates. The death-rate of a town for any particular year, as quoted by the Medical Officer of Health, is the *number of deaths per thousand population*. It is usually the *crude death-rate* because it does not take into account the age distribution of the population. In a town like Bournemouth, where a large number of elderly retired people reside, the crude death-rate is bound to be high compared with that of a newly developed industrial town whose population usually consists of young married couples and their children. In order to make a fair comparison of the death-rates of all types of town the *standardized* (or *corrected*) *death-rate* is used. This is a weighted average in which the weights bear a definite relation to the age distribution of the population of the whole country. Table 9F shows its method of calculation. In the first two columns the population of the town is classified according to age, and in the third column the number of deaths in each age group is recorded. The crude death-rate of the town is simply its total number of deaths divided by its total population in thousands. In the fourth column we have the STANDARD age distribution, in this case the percentage of the total population of the whole country in each age group. From the second and third columns the death-rate of each age group is calculated and recorded in the fifth column. These are often called the AGE-SPECIFIC DEATH-RATES. The standardized death-rate is the weighted average of the age-specific death-rates using the standard age distribution as weights.

TABLE 9F
Calculation of the crude and standardized death-rate

Age group	Population p	Deaths in age group d	Standard % of total population of Great Britain in each age group S	Age-specific death-rate. Death-rate of each age group $1000d/p$ R	SR
0–5	1000	13	12·1	13·00	157·30
5–15	1250	1	17·7	0·80	14·16
15–25	600	2	15·8	3·33	52·61
25–35	900	3	15·1	3·33	50·28
35–45	800	3	14·0	3·75	52·50
45–55	750	8	11·6	10·67	123·77
55–65	500	12	7·7	24·00	184·80
Over 65	400	20	6·0	50·00	300·00
Total	6200	62	100·0	—	935·42

$$\text{Crude death-rate} = \frac{\text{Total number of deaths}}{\text{Total population in thousands}}$$

= 10 (deaths per thousand population).

Standardized death-rate = $\Sigma SR/\Sigma S$

= 9·35 (deaths per thousand population).

83

In exercise 8, §101, the death-rate in 1951 is compared with the death-rate in 1931 by standardizing the 1951 death-rate by taking the age distribution in 1931 as the standard.

101. Exercises.

1. A composite index number is to be constructed from the following index numbers weighted as shown. Using a working mean of 140, or otherwise, calculate the composite index number as an arithmetic mean:

Index numbers	172	166	150	135	130
Weights	1	2	4	6	3

(London.)

2. What is meant by 'weighting' in the construction of index numbers?
Using a working mean of 115, or otherwise, calculate as an arithmetic mean correct to the nearest whole number the 'interim index of retail prices' for 18 April 1950, using the following table of price relatives and weights:

	Price relatives	Weights
Food	122·0	348
Rent and rates	101·3	88
Clothing	118·4	97
Fuel and light	115·2	65
Household durable goods	110·6	71
Miscellaneous goods	113·3	35
Services	106·6	79
Drink and tobacco	107·5	217

(London.)

3. Explain what is meant by an index number.
Calculate as an arithmetic mean correct to the nearest integer a cost-of-living index from the following table of price relatives and weights:

	Price relative	Weight
Food	122	35
Rent	101	9
Clothing	118	10
Fuel	115	7
Miscellaneous	108	39

(London.)

4. The following table gives the index numbers of three commodities in 1950, taking 1948 as base:

	1948	1950
Food	100	111
Fuel and light	100	105
Clothing	100	106

(a) Calculate the weighted average of these index numbers for 1950, when food, fuel and light, and clothing are given weights of 5, 1 and 2 respectively.
(b) Calculate, each to the nearest whole number, the 1948 index numbers referred to 1950 as base.

(London.)

5. Explain what is meant by a weighted mean. Why is it used instead of a simple mean?

A candidate obtains the following percentage marks in an examination:

Mathematics 72, French 52, History 67, English 59, Physics 82, Geography 66, Chemistry 77.

It is agreed to give treble weight to Mathematics, English and French, and double weight to Physics and Chemistry. What is the candidates weighted mean percentage? (London.)

6.　　　　*Interim index of retail prices,* 15 *June* 1954

	Price index (15 Jan. 1952= 100)	Weight
Food	113·6	399
Rent and rates	111·9	72
Clothing	96·3	98
Fuel and light	110·0	66
Household durable goods	95·2	62
Miscellaneous goods	100·1	44
Services	110·1	91
Alcoholic drink	101·4	78
Tobacco	100·3	90

Using the given weights, calculate the weighted arithmetic mean of the above price indices. (Northern.)

7. Distinguish between the crude and the standardized death-rates. Calculate their values from the following table, in which the last row gives the assumed standard age distribution per thousand:

Age group	0–9	10–39	40–59	60–
Population	20,000	60,000	15,000	5,000
Deaths in 1953	250	300	350	400
Standard distribution	250	500	200	50

(London.)

8.　　　*Population and deaths of infants under* 5 *years old* (*England and Wales*) (*in thousands*)

Year	Under 2 years old		2 years and under 5 years		Total	
	Population	Deaths	Population	Deaths	Population	Deaths
1931	1195	48·9	1796	8·4	2991	57·3
1951	1367	22·0	2355	2·4	3722	24·4

Calculate crude death-rates for the infant population in both years. Calculate age-specific death-rates, and hence the standardized death-rate for 1951, taking the infant population of 1931 as standard.

Describe in words the most prominent features of the data revealed by your analysis. (Northern.)

10

Miscellaneous Topics

102. Diagrammatic representation of statistical data. The student is already familiar with the construction and use of histograms, cumulative frequency curves, scatter diagrams and historigrams. In this chapter various other methods of presenting statistical data in pictorial or diagrammatic forms will be described and discussed.

103. The picturegram or pictogram. *The diagram below is an attempt to show in pictorial form the relative amounts of wheat obtained from different sources and consumed in the United Kingdom in a certain year. The actual amounts are in the ratio* 7:2:1.

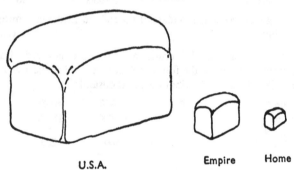

U.S.A.   Empire   Home

State whether you consider that the diagram gives a clear indication of this relation, giving reasons for your opinion. (London.)

The above example of a *picturegram* or *pictogram* does not give a clear indication of the given relation. It is very misleading because the loaves have been drawn as similar solids with *lengths* in the ratio 7:2:1. This means that their *volumes* would actually be in the ratio $7^3:2^3:1^3$, which is 343:8:1. The eye is better trained to compare lengths than volumes and a comparison of these volumes is even more difficult to assess because they are drawn on a flat sheet of paper. A true representation of the above data would be as shown on p. 87.

The popular newspapers and various forms of advertisements provide an infinite variety of pictograms in which colour as well as shape is used with great effect.

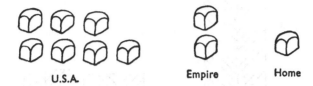

U.S.A.   Empire   Home

104. Bar diagrams. *Numbers of workers in certain industries in England and Wales, 1951 (in thousands):*

	Males	Females	Total
Textiles	401	476	877
Leather, leather goods and fur	44	26	70
Clothing	215	435	650
Food, drink and tobacco	419	240	659

Calculate percentages to compare the number of males in each industry with the corresponding number of females and present them in a self-explanatory table.

Construct a bar diagram to display simultaneously a comparison between the total numbers of workers in each industry and the sex composition of each total. (Northern.)

Simple arithmetic shows that 401 and 476 are respectively 45·7 and 54·3 % of 877, and table 10A gives the complete list of required percentages. Fig. 20 shows the data presented as a *bar diagram* in which, for example, the rectangle which represents the 877,000 textile workers is divided in the ratio 45·7:54·3 to show the sex composition of the industry. Fig. 21 is an alternative diagram more like a bar and less like a histogram than fig. 20.

TABLE IOA

Industry	Percentage of male workers	Percentage of female workers
Textiles	45·7	54·3
Leather, leather goods and fur	62·9	37·1
Clothing	33·1	66·9
Food, drink and tobacco	63·6	36·4

105. Circular diagrams. Table 10B, originally published in *The Times* on 19 March 1956, shows the National Savings receipts for the weeks ending 10 and 17 March 1956. Fig. 22 shows these receipts represented by two circles whose areas are proportional to the total receipts. Thus:

$$\frac{\text{The area of the circle representing the receipts for 17 March}}{\text{The area of the circle representing the receipts for 10 March}} = \frac{30\cdot230}{24\cdot496}.$$

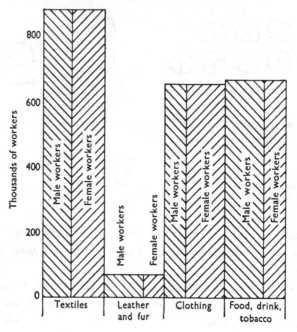

Fig. 20. Bar diagram.

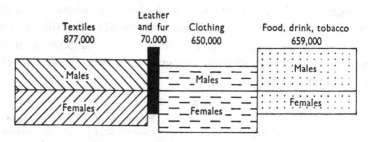

Fig. 21. Alternative form of bar diagram.

This means that $\dfrac{\pi R^2}{\pi r^2} = \dfrac{30 \cdot 230}{24 \cdot 496}$, where R and r are the radii of the two circles.

Hence
$$\frac{R}{r} = \sqrt{\frac{30 \cdot 230}{24 \cdot 496}} = 1 \cdot 111,$$

and in fig. 22 the radii of the two circles are actually 1 and 1·11 in.

Further, the first circle is divided into sectors containing angles

88

National Savings receipts (£ million)

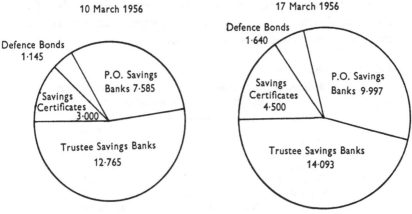

Fig. 22. Circular or pie diagrams.

proportional to 3·000, 1·145, 7·585 and 12·765. This means that the 360 degrees of the circle have to be divided in the ratio

$$3 \cdot 000 : 1 \cdot 145 : 7 \cdot 585 : 12 \cdot 765,$$

and hence the sector representing Savings Certificates contains an angle of $(3 \cdot 000 / 24 \cdot 496) \times 360$ degrees, that representing Defence Bonds contains an angle of $(1 \cdot 145 / 24 \cdot 496) \times 360$ degrees and so on. The angles of each sector are shown in table 10C.

TABLE 10B
National Savings receipts (£ million)

	Week ending	
	10 March 1956	17 March 1956
Savings Certificates	3·000	4·500
Defence Bonds	1·145	1·640
P.O. Savings Banks	7·585	9·997
Trustee Savings Banks	12·765	14·093
Totals	24·495	30·230

TABLE 10C
The angles of the sectors of the circular diagrams

Savings Certificates	44	53½
Defence Bonds	17	19½
P.O. Savings Banks	111	119
Trustee Savings Banks	188	168
Totals	360	360

Circular diagrams such as these are often called, for an obvious reason, *pie diagrams*. They are admirable for comparing totals and for displaying also the separate constituents of the totals but, arithmetically, they are more involved than bar diagrams.

106. Exercises.

1.   *Average expenditure of industrial and agricultural households in January* 1941 (*per household*)

		Industrial £ s. d.			Agricultural £ s. d.	
Food		1	14	1	1	8	9
Rent and rates			10	10		4	9
Clothes			8	1		5	3
Fuel and light			6	5		4	11
Other items		1	5	7		14	8
	Total	£4	5	0	£2	18	4

A pictogram is to be made of the above distributions suitable for publication in a periodical, the elements of the pictogram being symbols representing coins. Show, by a rough sketch, how you think the pictogram should be constructed.   (Cambridge.)

2. The following figures show the number of boys and girls leaving the Sixth Form Science in a random sample of 130 grammar schools, classified according to their future education or employment.

Leaving for teachers' training colleges	138
Leaving for technical schools	141
Leaving for universities	635
Leaving to take up paid employment	329
Total	1243

                                                                    (Cambridge.)

Present the data (i) as a bar diagram, (ii) as a circular diagram.

3. In government and other publications, statistical information is often shown in the form of ideographs; for example, pictures of sacks of different sizes, or alternatively pictures of numbers of sacks of the same size, may be used to show the amounts of wheat and flour obtained from different countries.

Comment on the advantages and disadvantages of these pictorial systems.
                                                                       (London.)

4. What rules should be observed when diagrams are used to present statistical data?

State which diagrams you would use to illustrate the following kinds of numerical data, giving brief reasons for your choice.

(i) The number of births in a borough for each month of the year 1952.

(ii) The total harvest yields of the four main cereal crops, wheat, barley, oats and rye in the United Kingdom for the year 1952.

(iii) The number of pairs of socks of different sizes issued to an infantry unit during the year 1952.   (London.)

5. Consumers' expenditure on food in the United Kingdom in 1953 is shown below:

	Millions of £	
Bread and cereals	527	
Meat and bacon	740	
Fish	93	
Oils and fats	156	
Sugar, preserves and confectionery	358	
Dairy products	521	
Fruit	232	
Vegetables	301	
Beverages	145	
Other manufactured foods	90	(London.)

Exhibit these facts in a suitable diagram.

6. The following table gives the production of steel in five districts of Great Britain in 1953:

District	A	B	C	D	E
Thousand tons	800	1350	2400	1850	3600

Construct a suitable bar chart and also a pie chart of radius 2 in. to illustrate these figures. (London.)

7. *University students classified according to sex and living quarters, 1946–7*

	Men	Women	Total
Colleges and hostels	10,800	6,100	16,900
Lodgings	19,000	5,400	24,400
At home	19,900	7,100	27,000

Calculate percentages to compare the distribution of men students among the three types of accommodation with the corresponding distribution of women students, and present them in a self-explanatory table.

Construct a bar diagram to display simultaneously a comparison between the total numbers of students living in each type of accommodation and the sex composition of each total. (Northern.)

8. (a) During 1954 each 29s. 0d. collected from Leeds ratepayers was allocated to the various services in the following way: Education 7s. 6d., Health 7s. 6d., Highways 4s. 0d., Police and Fire Brigade 2s. 6d., Street Lighting 1s. 6d., Other Services 3s. 6d. and Administration 2s. 6d. Construct a diagram which illustrates these data. (Northern.)

9.

Customs and Excise, 1955–6 (£ millions)	
Purchase tax	419
Tobacco	$668\frac{1}{2}$
Beer	258
Spirits	125
Light wines	$20\frac{1}{2}$
Entertainments	$39\frac{1}{2}$
Football pools	$20\frac{1}{2}$
Total	1551

Present the above data (i) as a bar diagram, (ii) as a circular diagram.

10. *Jap shipyards grab lead*

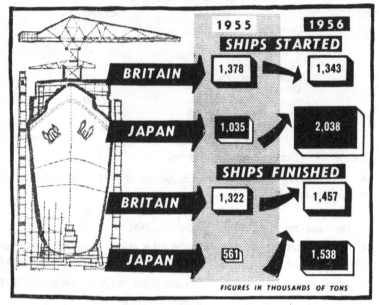

1955   1956

SHIPS STARTED

BRITAIN   1,378   1,343

JAPAN   1,035   2,038

SHIPS FINISHED

BRITAIN   1,322   1,457

JAPAN   561   1,538

FIGURES IN THOUSANDS OF TONS

Criticize the above 'Expressograph' (by Michael Rand) which appeared in the *Daily Express* on 23 January 1957.

Illustrate the data by a simple bar diagram.

Since the number of ships *finished* is of greater importance than the number of ships *started*, do you think that the caption 'Jap shipyards grab lead' is fully justified?

107. Sampling. A farmer buying a large consignment of grain might be seen to roll up his coat sleeve, thrust his hand well into the middle of one of the sacks and draw forth a handful of the contents for careful examination. After repeating this process with three or four more sacks taken at random the farmer's opinion of the quality of the whole consignment of grain would be established. This idea of estimating the properties of the whole population by investigating a *random sample* taken from the population is widely used in statistical research. In production engineering, for example, *quality control* depends on the estimation of the variability of the complete output by examining the variability of random samples of the output. Examples 1 and 2 of §46 are samples from which an estimate of the whole output can be made.

108. Bias. Some years ago, before the days of Market Research, a manufacturer of porridge oats wished to know what percentage of the population of London took porridge for breakfast. He thought that a quick estimate might be obtained by opening at random the telephone directory to any page and inquiring tactfully of all the subscribers on that page if they had taken porridge for breakfast that morning. He was astonished to find that all the subscribers questioned in this way had actually taken porridge that morning until he realized that the directory had been opened, quite by chance, at a page on which the initial letter of the surnames was *M* and the surnames were actually McPherson, Macquire, McRae, etc. This is an example of an extremely *biased* sample. Such a sample gives a completely erroneous conclusion about the whole population.

A second example of bias is that of an inquiry made by a certain Medical Officer of Health who wished to find out what proportion of the old people of Great Britain are still at work after the age of 65. He sent our questionnaires to about 11,000 Darby and Joan Clubs, and replies showed that a very minute percentage indeed of the Members of these clubs, who were over 65, were still at work. The Medical Officer's conclusion that most people retire at 65 was, of course, quite wrong, because only old people who *are* retired have time to join the clubs, so that his sample was extremely biased.

109. Random sampling. The two examples of bias given in §108 emphasize the fact that the results obtained from a sample must not be applied to the whole population unless the sample is random and fully representative of the whole field of inquiry. *A random sample is one for which every member of the group has an equal chance of selection.* To draw a random sample of names from the telephone directory, the porridge manufacturer might have used two sets of cards, the first bearing the numbers of all the pages in the directory and the second bearing numbers equal to the number of names on each page. After thoroughly shuffling the two sets of cards separately, the number 273, for example, drawn at random from the first set with, say, 94 drawn at random from the second set, would indicate that the 94th name on p. 273 had to be contacted. By repeating this process 500 times a random sample would be drawn from the telephone directory, but it would not be representative of the whole population of London. It would only represent that section (or stratum) of the population which could afford telephones.

It should be noted that the tedious process of mixing and drawing out numbered cards can be avoided by using the table of Random Sampling Numbers constructed by L. H. C. Tippett.

If the porridge manufacturer had been determined to get a truly representative random sample he could have drawn a sample of addresses from the Electoral Register. This would have involved a great deal of office work because the Register is kept in the form of a large number of booklets, one for each polling district. When the random sample of addresses had been drawn, interviewers would have had to be sent to these addresses. (The sending of interviewers is much more satisfactory than the sending of questionnaires requiring a written postal reply because random sampling demands that every effort be made to obtain the necessary information from the selected addresses.) Also, without doubt, the interviewers would have had to make repeated calls at many of the addresses.

The above description shows that, although random sampling is a long expensive operation, it does give a reliable picture of the whole population.

110. Quota sampling. A cheaper and quicker method of sampling than that described in §109 is *quota sampling*. It is much used in social and market surveys. In this type of sampling each interviewer is given definite instructions about the section of the public he is to question, but the final choice of the actual persons is left to his own convenience and is not predetermined by some carefully operated randomizing plan. Each interviewer might be told, for example, to question 50 people who are to include at least 5 men over 40 years of age, 5 men between 20 and 40, 5 men or boys under 20, 5 schoolboys, 5 men dressed in lounge suits, 5 men dressed in boiler-suits, overalls or working clothes. Similarly, precise instructions would be given about women and girls. It will be realized, of course, that some of the men in lounge suits will also be in the over 40 class, so that an interviewer might actually cover the first two classes and the last two by interviewing only 10 men for the four classes. The interviewers are stationed throughout Great Britain at various places in town and country. They make their interviews during the course of a morning, telephone the results to a central office and a final announcement about the survey can be made 24–48 hr. after its commencement. Quota sampling is relatively inexpensive to operate, and in skilled hands can give good results. It has its pitfalls, however, as the following example shows. A survey was made a year or two ago to find out what percentage of the population of Great Britain took part in gambling. One interviewer reported the astonishing news that all 50 people of his quota were regular gamblers. Subsequent inquiries revealed that he had been interviewing in a street which led to a railway station from which excursion trains to Newmarket Races were leaving at frequent intervals on that particular morning.

111. The framing of questions and questionnaires. It is very important that questions put to members of the general public by interviewers (or in questionnaires) should be capable of a precise answer. 'Have you had a serious illness recently?' would be unsatisfactory. Instead, the interviewer might ask:

(i) 'Have you been under the doctor during the last twelve months?' The answer 'Yes' being followed by:

(ii) 'How long did the doctor attend you?' and

(iii) 'Did you have to be taken to hospital?'

Again, 'What sort of education have you had?' would be thoroughly unsound. Instead, several questions should be asked such as:

(i) 'What schools did you attend?'

(ii) 'Have you been to a College or University?'

(iii) 'What examinations have you passed?'

In general, questions should be put simply and straightforwardly so that they are easily understood by the unintelligent and uneducated. They should be capable of being answered by 'Yes', 'No', a number, a place, a date or something equally precise. They should be framed tactfully so as to break down any barrier of suspicion or reserve on the part of the informant. (An interviewer is so much more successful than a questionnaire in this respect.) Finally, ambiguity in questions should be avoided at all costs.

11

Miscellaneous Problems

1. Describe briefly a statistical investigation in which you have taken part, referring particularly to
(a) preliminary planning,
(b) the methods used to collect and tabulate data,
(c) the methods used to analyse the data, and
(d) any conclusions reached. (London.)

2. The following table gives details of weekly income and of weekly expenditure on food of a group of four families:

Family	Total family income (£)	Expenditure per head on food (£)	Total number in family
A	14·8	1·6	4
B	28·2	1·8	6
C	16·4	1·4	5
D	35·2	2·1	4

Find to the nearest shilling the average income per head of the group and the average expenditure on food per family.

Which families spend (a) more, (b) less than the average on food and by how much? (London.)

3. The numbers of thousands of live births registered in Wales, Scotland and Northern Ireland in 1954 were:

	Wales	Scotland	Northern Ireland	Total
1st quarter	10·4	23·5	7·3	—
2nd quarter	10·1	24·1	7·7	—
3rd quarter	10·0	22·8	7·1	—
4th quarter	9·4	21·9	6·8	—

Find for each quarter the total of the numbers registered.

Construct a bar diagram to illustrate simultaneously these totals and the contributions to these totals from each country. (London.)

4. Explain carefully the meaning of the following terms: (a) index number, (b) frequency distribution, (c) median, (d) mode, and (e) dispersion. (London.)

96

5. The following table shows the monthly production of cane sugar in units of 100 tons for the period July 1953 to June 1955:

	J	F	M	A	M	J	J	A	S	O	N	D
1953	—	—	—	—	—	—	22	27	24	21	18	16
1954	17	17	16	19	23	24	26	23	26	19	20	15
1955	17	19	19	25	27	28	—	—	—	—	—	—

Construct a table showing the twelve-monthly moving average correct to the nearest 10 tons and draw a graph to show the trend. (London.)

6. In a botanical experiment the lengths of 100 leaves from a certain species of plant were measured and the frequencies tabulated as follows:

Length of leaf in mm.	Frequency
20–24	1
25–29	5
30–34	10
35–39	19
40–44	25
45–49	21
50–54	15
55–59	3
60–64	1

Construct a histogram to show the frequency distribution and state the mode. (London.)

7. Using the frequency distribution of Question 6, with a working mean of 42, calculate the mean length of leaf and the standard deviation from the mean. (London.)

8. From the data of Question 6 construct a cumulative frequency table and draw the ogive. Use the ogive to determine (i) the median length of leaf, and (ii) the difference in length between the first and third quartiles. (London.)

9. The percentage marks for ten candidates for Intelligence and Arithmetic tests were as follows:

Candidate ...	A	B	C	D	E	F	G	H	J	K
Intelligence test	27	39	48	42	47	55	35	30	32	45
Arithmetic test	40	44	65	67	80	84	60	57	46	55

Plot a scatter diagram and a line of best fit.

A candidate X scored 40 in the Intelligence test and a candidate Y scored 55 in the Arithmetic test. Each was absent from the other test. Use your diagram to estimate a comparison between the performances of X and Y. (London.)

10. Arrange the table of results in Question 9 according to ranks and calculate the rank correlation coefficient.

State what significance, if any, you would attach to the result.

(London.)

11. Give an example to illustrate the construction and use of each of the following: (i) histogram or column diagram, (ii) pie chart, (iii) cumulative frequency diagram. (London.)

12. The total area sown with oats in Great Britain in 1955 was 2,581,000 acres, *to the nearest* 1000 *acres*. The average yield was 22 cwt per acre, *correct to the nearest cwt*. Estimate the total yield of oats in tons as accurately as the data allow.

Calculate also, in tons, the extreme limits between which the total yield must lie. (London.)

13. Calculate the mean and standard deviation of the following set of numbers:

$$3, \quad 4, \quad 7, \quad 12, \quad 8, \quad 9, \quad 5, \quad 16. \qquad \text{(London.)}$$

14. The following table shows the coal production (in millions of metric tons) of (*a*) United States of America, (*b*) Great Britain, and (*c*) Germany, for the first half of the twentieth century:

Years	(*a*)	(*b*)	(*c*)
1900–04	300	200	190
1905–09	400	250	200
1910–14	450	300	250
1915–19	540	250	220
1920–24	530	230	230
1925–29	540	220	300
1930–34	400	210	270
1935–39	400	230	350
1940–44	550	200	440
1945–49	550	200	230

Construct a histogram to illustrate both the relative and the combined production of the three countries. (London.)

15. The following pairs of values of *x* and *y* were obtained by experiment:

x	0·5	1·1	1·8	2·2	2·7	3·2	3·8	4·7
y	2·3	2·5	2·9	3·2	3·4	3·5	4·0	4·3

Plot these results in a diagram and draw a line of best fit. Use your graph to estimate (i) the value of *y* when *x* = 2·5, and (ii) the value of *x* when *y* = 5·0. (London.)

16. The index numbers of commodities A, B and C in 1956, taking 1939 as base are shown below:

	1939	1956
A	100	186
B	100	242
C	100	282

Calculate the weighted average of these index numbers for 1956 when A, B and C are given weights of 4, 2 and 3 respectively.

Calculate, each to the nearest whole number, the 1939 index numbers referred to 1956 as base. (London.)

17. The following table was drawn up from the height measurements of 100 recruits:

Height in inches (mid-point of interval)	62	64	66	68	70	72
Frequency	3	12	32	24	21	8

Construct a cumulative frequency table and draw the ogive. Use the ogive to determine the median height.

What is the modal height of the distribution? (London.)

18. Using the frequency distribution of Question 17, with a working mean of 66 inches, find the mean height and the standard deviation from the mean. (London.)

19. The following table shows estimated personal expenditure (in £1,000,000) in Great Britain on fuel and light:

	Quarter			
	(1)	(2)	(3)	(4)
1953	133	106	92	116
1954	144	116	100	125
1955	155	121	106	139
1956	175	—	—	—

Calculate the four-quarterly moving average and draw a graph to show the trend. (London.)

20. The marks of 10 students in Arithmetic and Algebra tests were:

Student ...	A	B	C	D	E	F	G	H	I	J
Arithmetic	49	60	41	34	21	42	43	65	45	63
Algebra	39	74	33	32	13	31	34	71	57	40

Calculate the coefficient of rank correlation between the performances in Arithmetic and Algebra. (London.)

21. *Weight of litter left by visitors in London parks on August Bank Holiday 1955*

	Left in baskets (cwt.)	Left on ground (cwt.)
Hyde Park	61	21
Kensington Gardens	16	7
St James's and the Green Parks	24	6
Greenwich Park	7	5
Regent's Park	66	31
Richmond Park	8	7
Hampton Court	14	8
Bushey Park	4	5
Total in cwt.	200	90

Display the above data by two circular diagrams, the total weight of litter left in baskets being represented by a circle of radius $2\frac{1}{4}$ inches. All the necessary calculations should be shown clearly and the radius of the circle representing the total weight of litter left on the ground should be stated. (Northern.)

22. (a) Four numbers a, b, c, d are such that b is 90% of a, c is 80% of b and d is 120% of c.

Express (i) c as a percentage of a, (ii) d as a percentage of a.

(b) A form consists of n_1 boys and n_2 girls. In an examination the average mark of the boys was x and the average mark of the girls was y. Find the average mark of the whole form. (Northern.)

23. *Weights, in lb., of crews taking part in the finals at the Henley Royal Regatta 1956*

	Bow	2	3	4	5	6	7	Stroke	Cox
Royal Air Force	160	171	176	185	165	168	166	157	138
Princeton University, U.S.A.	154	155	159	171	173	160	152	152	126
Peterhouse, Cambridge	174	172	175	168	188	179	161	171	126
Magdalene College, Cambridge	176	172	166	183	172	170	155	155	117
St Paul's School	162	155	171	159	188	181	163	166	112
Eton College	153	170	176	170	189	177	168	188	116
L'Armée Française	173	169	175	185	227	190	185	167	112
Roddklubben, Sweden	171	186	164	180	192	189	180	173	122

Form a frequency distribution for the above weights of 72 men and construct the histogram. The following grouping is appropriate: 109·5–119·5, 119·5–129·5, 129·5–139·5, etc.

Comment on any special features of the histogram.   (Northern.)

24. The first three rows of the following table show extracts from the vital statistics of a certain town while the last row gives an assumed standard age distribution per thousand:

Age group ...	0–9	10–39	40–59	60–
Population in thousands	16	48	12	4
Deaths in 1955	212	247	291	337
Standard age distribution	150	400	300	150

Calculate the crude and standardized death-rates and explain why crude death-rates are considered inadequate.   (Northern.)

25. Describe any statistical investigation in which you have assisted (or would like to assist). The description should state clearly
   (i)   the object of the investigation,
   (ii) the methods of obtaining information,
   (iii) the precautions necessary for avoiding bias,
   (iv) the method of tabulating the results.   (Northern.)

26. The following table shows the frequency distribution of the examination marks of 700 candidates in a Mathematics examination:

Examination marks	Number of candidates	Examination marks	Number of candidates
10–19	10	60–69	148
20–29	32	70–79	127
30–39	45	80–89	90
40–49	87	90–99	44
50–59	117		

Taking a working zero in the class 60–69 and a unit of 10 marks calculate the arithmetic mean.

Construct a cumulative frequency curve and use it to determine:
(i) the percentage of candidates passing if the pass mark is 55,
(ii) the lowest mark for distinction if 5% of the candidates are to be given distinction.   (Northern.)

27.   *Cinemas in the East and West Ridings. Analysis
by size, 27 March 1954*

Seating capacity	Number of cinemas
Under 251	3
251– 500	70
501– 750	127
751–1000	125
1001–1250	56
1251–1500	25
1501–1750	17
1751–2000	7
2001–2250	5
2251–2500	4
2501–2750	3
2751–3000	2
Total	444

Taking a working zero in the class 751–1000 and a unit of 250 seats, calculate the arithmetic mean seating capacity of the cinemas.

Construct a cumulative frequency curve and use it to determine the median.   (Northern.)

28. (*a*) The percentage of teachers who are men is *a*, the percentage of teachers who have honours degrees is *b*, and the percentage of teachers with honours degrees who are men is *c*. Deduce the percentage of men teachers who have honours degrees.

(*b*) In a survey designed to ascertain the views of grammar school pupils the following questions are put to samples of boys from grammar schools, the samples being selected by the headmasters:

(i) *Do you think that the more modern subjects, such as Physics, Chemistry, and Biology, are more important than the traditional subjects, such as History, Latin and Greek?*

(ii) *Do you intend to stay on at school after the minimum leaving age permitted by law?*

Criticize the questions and the method of selecting the sample from the point of view of potential bias.   (Northern.)

29. The following, in chronological order, were the annual numbers of injuries (in units of a thousand) suffered by colliery workers for the years 1924–51 inclusive: 195, 178, 91, 173, 162, 176, 166, 141, 126, 122, 133, 134, 136, 141, 132, 134, 146, 158, 167, 174, 177, 181, 167, 163, 183, 229, 238, 234.

Taking class intervals 90–, 120–, 150–, 180–, and 210–240, construct a frequency table for the data, and draw a histogram. Point out its most prominent features.

Discuss briefly whether the histogram alone is a fair summary of the data, and whether it would be a fair conclusion that before 1948 the mines were rarely as dangerous as they were during the period 1948–51.

(Northern.)

30.   *Frequency of instantaneous speeds of motor vehicles*
         at a point in a busy London street

Speed (m.p.h.)	Percentage frequency	Speed (m.p.h.)	Percentage frequency
0–	0	20–	25·5
5–	5·5	25–	9·0
10–	15·0	30–	3·0
15–	41·5	35–40	0·5

Taking a working zero in the class 15– and a unit of 5 m.p.h., calculate the arithmetic mean speed, using the frequency table above. Construct a cumulative percentage frequency graph, and use it to estimate the 10–90 percentile range. (Northern.)

Answers to Exercises

§13. 1. 0–29. **2.** 169½–179½. **3.** 50–59; 63.

§15. 1. 148½ lb. **2.** 48 in.

§26. 1. 53, 70, 32½. **2.** 25·3, 9·25, 2·72.
 3. 48·8 in., 57·5 in., 40·2 in. **4.** 155 lb., 169 lb., 143 lb.
 5. 328; 315; 362, 246. **6.** 15, 56; 71, 31.
 7. 70, 84, 46; 63, 73. **8.** 45; 58; 59, 31.
 9. 45s. 6d., 52s. 6d., 37s. 6d. **10.** 33·2, 34·8, 31·0.
 11. 35·7, 37·0, 34·2; 67·5, 69·3, 65·7 in. **12.** 64200, 95.

§28. 1. 981·2. **2.** 42·4. **3.** 100·1.

§31. 2. 0·55. **3.** 2.

§35. 3. 156·1 lb. **4.** 48·8 in. **5.** 21 min. **6.** 302·7.
 7. 44s. 11d. **8.** 35·7 in., 67·5 in.

§44. 1. 222⅔ m.p.h. **2.** 37 1/19 m.p.g.

§51. 1. 0·025 in., 0·009 in. **2.** 0·63. **3.** 7·6. **4.** 100·2, 1·03.
 5. 9·1, 4·8. **6.** 6·71, 9·93.

§53. 1. 68·63. **2.** 12·45, 13·05. **3.** 22·0. **4.** 8s. 3d.
 5. 2·36.

§58. 1. 0·74. **2.** 10·25. **3.** 1·35. **4.** 10·9, 5·66.
 5. 7·72, 11·97.

§63. 1. 88·2. **2.** 15·1, 16·2. **3.** 10s. 11d. **4.** 3·26.
 5. 0·057, 0·039. **6.** 16·7 %, 18 %. **7.** 15·6 %, 20 %.

§74. 1. French and German marks are directly correlated. German marks
 about 20 better than French.
 2. (*a*) Botany and Zoology marks are directly correlated.
 (*b*) Marks are about equal in each subject.
 3. 19·3 in.
 4. (*a*) Physics marks about 15 better than Mathematics.
 (*b*) Physics and Mathematics marks directly correlated.
 5. 0·8. **6.** 1·7. **8.** *O* is farthest above the line.
 9. 12·5; 11·4 along 10·8 across. **11.** 26, 33, 43, 52, 61, 71; 58 years.
 12. 77, 77, 85, 95, 113, 127, 141, 144; 127 lb.

§80. 1. 0·98. **2.** 0·98. **3.** Unity. **4.** 0·97.
 5. 0·32. **6.** 0·86. **7.** −0·93. **8.** 0·83.

9. 0·86.   **10.** 0·82.   **11.** 0·91.   **12.** 0·34.

13. 0·98.   **14.** 0·89.   **15.** 0·92.

§84. **1.** 0·96.   **2.** 0·98.   **3.** Unity.   **4.** 0·90.

5. 0·26.   **6.** 0·89.   **7.** −0·95.   **8.** 0·82.

9. 0·81.   **10.** 0·80.   **11.** 0·91.   **12.** 0·37.

13. 0·997.   **14.** 0·98.   **15.** 0·95.   **16.** 0·78.

17. 0·46.   **18.** 0·8.   **19.** 0·02.   **20.** 0·62.

§88. **1.** $73\frac{3}{4}$, $74\frac{1}{2}$, $75\frac{1}{4}$, $75\frac{1}{2}$, $76\frac{1}{2}$, $77\frac{3}{4}$, $79\frac{1}{2}$, 81, $82\frac{1}{4}$.

2. 86·4, 86·525, 89·175, 90·425, 91·575, 92·525, 90·0, 88·125, 87·025.

3. 11·275, 11·075, 11·925, 12·425, 13·050, 13·175, 11·800, 10·625, 9·700, 9·575, 9·825, 10·375, 10·850.

4. $24\frac{1}{4}$, 25, 25, 28, $30\frac{3}{4}$, $33\frac{1}{2}$, $37\frac{1}{2}$, $39\frac{1}{4}$, 38.

5. $72\frac{3}{4}$, $74\frac{3}{4}$, $76\frac{1}{4}$, $76\frac{3}{4}$, $77\frac{1}{2}$, $79\frac{1}{2}$, $82\frac{1}{4}$, $84\frac{1}{2}$, $87\frac{1}{4}$.

6. $323\frac{1}{4}$, $314\frac{3}{4}$, $311\frac{1}{4}$, $316\frac{1}{2}$, $327\frac{1}{4}$, $349\frac{1}{2}$, $367\frac{1}{4}$, $376\frac{1}{2}$, $385\frac{3}{4}$, 394, $402\frac{1}{2}$, $409\frac{1}{2}$, $416\frac{1}{4}$, $418\frac{3}{4}$, 423, $428\frac{1}{4}$, $432\frac{1}{2}$, $438\frac{1}{4}$, $444\frac{3}{4}$, 451, $457\frac{1}{2}$.

§93. **1.** $20\frac{1}{12}$, $19\frac{11}{12}$, 20, $20\frac{1}{12}$, $19\frac{11}{12}$, $20\frac{1}{4}$, $20\frac{1}{3}$, $20\frac{5}{6}$, $21\frac{1}{4}$, $22\frac{1}{2}$, $23\frac{2}{3}$, $24\frac{5}{6}$, $26\frac{1}{12}$, 27, $28\frac{1}{12}$, $29\frac{1}{4}$.

2. (In hundreds) 175, $178\frac{7}{12}$, 181, 186, 190, $194\frac{7}{12}$, $199\frac{1}{2}$, $204\frac{3}{4}$, $212\frac{1}{4}$.

3. 17·5, 19·2, 21·2, 23, 25, 26·5, 28·2, 29·5, 30·5, 32, 33·2, 34·8, 35·8, 37·5, 39, 39, 39·3, 40, 40, 39·7, 39·5, 39·5, 39·5, 39·5, 39·5, 38·5, 37·5, 36·5, 35·2, 33·5, 32·8.

7. 81·9, 83·9, 85·4, 86·6, 87·5, 88·9, 89·7, 91·7, 93·1, 94·1, 95·2.

8. 73·8, 72·4, 71·0, 70·0, 67·4, 64·8, 61·0, 58·4, 57·6, 57·8 (beef), 92·4, 89·6, 87·6, 83·0, 78·4, 75·4, 72·0, 70·8, 73·8, 72·4 (mutton).

§101. **1.** 144.   **2.** 114.   **3.** 114.

4. (a) 109. (b) 90, 95, 94.   **5.** $66\frac{2}{3}$.   **6.** 107·3.

7. 13·0; 14·3.   **8.** 19·16, 6·56, 16·09; 7·04.

Chapter 11.

2. £5, £8. 3s. (a) B £2. 13s., D, 5s. (b) A £1. 15s., C £1. 3s.

5. 20·3, 20·7, 20·3, 20·5, 20·3, 20·5, 20·4, 20·6, 20·8, 21·3, 21·7, 22·0.

6. $42\frac{1}{2}$.   **7.** 42·3, 7·8.   **8.** (i) 43 mm.; (ii) 11 mm.

9. X better than Y.   **10.** 0·76.

12. 2,840,000 tons. Between 2,774,000 and 2,903,000 tons.

13. 8, 4·062.   **15.** (i) 3·26; (ii) 5·6.

16. $230\frac{4}{9}$; 54, 41, 35.   **17.** 67·2 in., 66·4 in.

18. 67·4 in., 2·5 in.

19. $111\frac{1}{4}$, $114\frac{1}{2}$, 117, 119, $121\frac{1}{4}$, 124, $125\frac{1}{2}$, $126\frac{3}{4}$, $130\frac{1}{4}$, $135\frac{1}{4}$.

20. 0·88.   **21.** $1\frac{1}{2}$ in.

22. (a) 72 %, 86·4 %. (b) $(n_1 x + n_2 y)/(n_1 + n_2)$.

24. 13·6, 24·0.   **26.** (i) 62·1; (ii) 65; (iii) 91.

27. 873, 796 seats.   **30.** 18·925 m.p.h.; 15 m.p.h.

Glossary of Terms used in this Work

Abscissa. The so-called 'horizontal' scale of a graph which is always used for the x-axis.

Absolute measure. A measure in terms of units.

Age-specific death-rate. The death-rate for a specified age-group.

Arithmetic mean. See 'mean'.

Average. A measure of central tendency of a group which enables us to assess the position in which the group stands with respect to other groups.

Bimodal. A frequency distribution with two modes.

Birth-rate. The number of live births in any year per thousand of the estimated population at the middle of the year.

Coefficient of correlation.

$$r_{xy} = \frac{\text{Covariance}}{\text{Product of Standard deviations}} = \frac{S_{xy}}{Sx . Sy}.$$

Coefficient of rank correlation.

$$R = 1 - \frac{6\Sigma D^2}{n(n^2 - 1)},$$

where D is the *rank difference* and n the *number of pairs*.

Coefficient of regression. The gradient of the regression line.

Coefficient of variation.

$$\text{Coefficient of variation} = \frac{\text{Standard deviation}}{\text{Mean}}$$

$$= \frac{100 \times \text{Standard deviation}}{\text{Mean}} \%.$$

Continuous variation. A distribution in which values of the frequency may exist at every point of the abscissa scale. It may be represented graphically as a continuous curve, i.e. a frequency curve.

Corrected death-rate. A weighted average of age-specific death-rates in which the weights bear a definite relation to the age distribution of the population of the whole country.

Crude death-rate. The number of deaths in any year per thousand of the population alive at the middle of the year.

Discrete variation. A distribution in which values of the frequency only exist at certain separate points of the abscissa scale. It may be represented by a frequency polygon, but should not be represented by a continuous curve.

Geometric mean. The geometric mean of the n values $x_1, x_2, x_3, ..., x_n$ is

$$\sqrt[n]{(x_1 . x_2 . x_3 ... x_n)}.$$

Harmonic mean. The harmonic mean of the n values $x_1, x_2, x_3, ..., x_n$ is

$$\frac{1}{\frac{1}{n}\left\{\frac{1}{x_1}+\frac{1}{x_2}+\frac{1}{x_3}+...+\frac{1}{x_n}\right\}}.$$

Infant mortality rate. The number of deaths of infants under one year of age per thousand live births.

Lower quartile. If n measurements are arranged in order of magnitude from the least to the greatest, the $\frac{1}{4}(n+1)$th is the lower quartile. The lower quartile divides the area of the histogram in the ratio $1:3$.

Mean. The arithmetic mean, or more simply the mean, of the n values $x_1, x_2, x_3, ..., x_n$ is

$$\frac{1}{n}\{x_1+x_2+x_3+...+x_n\}.$$

Mean deviation. Mean deviation $=\frac{1}{n}\Sigma\,|\,d\,|$, where $n=$ the number of values and $\Sigma\,|\,d\,|=$ the sum of the moduli of their deviations from the mean, median or mode.

Median. If n measurements are arranged in order of magnitude the $\frac{1}{2}(n+1)$th is the median. The median bisects the area of the histogram.

Mode. In a frequency distribution the value which occurs most frequently is called the mode (or norm).

Modulus. The modulus of a number is its numerical value, no regard being paid to its $+$ or $-$ sign.

Negatively skewed distribution. A distribution in which more than half of the area of the histogram is to the left (negative) side of the mode.

Normal frequency curve (or normal distribution). A symmetrical bell-shaped curve of fundamental importance in statistical theory. Its nature and significance will be dealt with in a 'A Second Course in Statistics'. Its mathematical equation for a frequency distribution of n measurements is

$$y=\frac{n}{\sqrt{(2\pi)}}\,e^{-\frac{1}{2}x^2},$$

provided the mean is transferred to the origin and the units are adjusted so that the standard deviation is unity. It is also known as the *error curve*, the *normal probability curve* or the *Gaussian curve*.

Ogive. The graph of a cumulative frequency distribution.

Ordinate. The so-called 'vertical' scale of a graph which is always used for the y-axis.

Parameters. A collective name given to statistical measures such as the mean, median, mode, mean deviation, standard deviation.

Percentiles. The values which divide the area of the histogram into a hundred equal parts.

Positively skewed distribution. A distribution in which more than half of the area of the histogram is to the right (positive) side of the mode.

Quartile deviation. Semi-interquartile range or quartile deviation

$$=\tfrac{1}{2}\,(\text{Upper quartile}-\text{Lower quartile}).$$

Quartiles. See 'Lower quartile' and 'Upper quartile'.

Relative measure. A ratio which is independent of units. A 'coefficient' is often, but not always a relative, measure.

Skewness. See 'positively skewed distribution' and 'negatively skewed distribution'.

Standard deviation. Standard deviation $= \sqrt{\left\{\dfrac{\Sigma d^2}{n}\right\}}$,

where $n =$ the number of values and $\Sigma d^2 =$ the sum of the squares of their deviations from the mean.

Standardized death-rate. See 'corrected death-rate'.

Upper quartile. If n measurements are arranged in order of magnitude from the least to the greatest, the $\frac{3}{4}(n+1)$th is the upper quartile. The upper quartile divides the area of the histogram in the ratio $3:1$.

Index

THE NUMBERS REFER TO PAGES

abscissa, 16, 106
absolute measure, 44, 106
age-specific death-rate, 83, 85, 106
arbitrary origin, 23, 40, 79
arithmetic mean, 23, 29, 30, 106
array of points, 50
averages, 28, 106

bar diagrams, 87
birth-rate, 13, 106
bivariate distribution, 46, 107

central tendency, 106
circular diagrams, 87
coefficient of correlation, 56, 106
coefficient of rank correlation, 62, 106
coefficient of regression, 49, 106
coefficient of variation 44, 106
composite price relative, 79
continuous variation, 2, 106
corrected death-rate, 83, 85, 106
correlation, 45, 56, 62
correlation by ranks, 62, 106
correlation coefficient, 56, 106
cost-of-living index, 80
covariance, 56
crude death-rate, 82, 106
cumulative frequency, 15
cyclical variations, 75

dependent variable, 51
direct correlation, 46
discrete variation, 2, 106
dispersion, 32, 42

four-quarterly moving averages, 66
frequency curve, 5
frequency distribution, 1
frequency polygon, 3

Galton graph, 56
general trend, 66, 74
geometric mean, 30, 106

harmonic mean, 30, 107

histogram, 1
historigram, 66

ideograph, 90
independent variable, 51
index numbers, 79
index of retail prices, 80
infant mortality, 13, 107
interim index of retail prices, 80
interpercentile range, 35
inverse correlation, 48

J-type distribution, 9

line of regression, 49
lower quartile, 19, 107

mean, 23, 29, 30, 107
mean deviation, 36, 107
mean of an array, 50
median, 18, 19, 107
mid-interval values, 25
minimum values of correlation coeffi-
    cient, 58
modal class (or group), 2
mode (or norm), 1, 29, 107
mode, determination of, 10, 30
modulus, 36, 107
moving averages, 66

negative skewness, 6, 9, 107
normal frequency curve, 5, 9, 107

ogive, 15, 107
ordinate, 16, 107

parameters, 107
percentage relative, 78
percentiles, 17, 107
picturegram (or pictogram), 86
pie diagrams, 90
positive skewness, 6, 10, 107
price index, 79
price relative, 78
product-moment, 57

o   109   LS

Tables

REPRINTED FROM

The Cambridge Four-Figure Mathematical Tables

LOGARITHMS OF NUMBERS

	0	1	2	3	4	5	6	7	8	9	1	2	3	4	5	6	7	8	9
											| Differences |								
10	0000	0043	0086	0128	0170	0212	0253	0294	0334	0374	4	8	12	17	21	25	29	33	3
11	0414	0453	0492	0531	0569	0607	0645	0682	0719	0755	4	8	11	15	19	23	26	30	3
12	0792	0828	0864	0899	0934	0969	1004	1038	1072	1106	3	7	10	14	17	21	24	28	3
13	1139	1173	1206	1239	1271	1303	1335	1367	1399	1430	3	6	10	13	16	19	23	26	2
14	1461	1492	1523	1553	1584	1614	1644	1673	1703	1732	3	6	9	12	15	18	21	24	2
15	1761	1790	1818	1847	1875	1903	1931	1959	1987	2014	3	6	8	11	14	17	20	22	2
16	2041	2068	2095	2122	2148	2175	2201	2227	2253	2279	3	5	8	11	13	16	18	21	2
17	2304	2330	2355	2380	2405	2430	2455	2480	2504	2529	2	5	7	10	12	15	17	20	2
18	2553	2577	2601	2625	2648	2672	2695	2718	2742	2765	2	5	7	9	12	14	16	19	2
19	2788	2810	2833	2856	2878	2900	2923	2945	2967	2989	2	4	7	9	11	13	16	18	2
20	3010	3032	3054	3075	3096	3118	3139	3160	3181	3201	2	4	6	8	11	13	15	17	1
21	3222	3243	3263	3284	3304	3324	3345	3365	3385	3404	2	4	6	8	10	12	14	16	1
22	3424	3444	3464	3483	3502	3522	3541	3560	3579	3598	2	4	6	8	10	12	14	15	1
23	3617	3636	3655	3674	3692	3711	3729	3747	3766	3784	2	4	6	7	9	11	13	15	1
24	3802	3820	3838	3856	3874	3892	3909	3927	3945	3962	2	4	5	7	9	11	12	14	1
25	3979	3997	4014	4031	4048	4065	4082	4099	4116	4133	2	3	5	7	9	10	12	14	1
26	4150	4166	4183	4200	4216	4232	4249	4265	4281	4298	2	3	5	7	8	10	11	13	1
27	4314	4330	4346	4362	4378	4393	4409	4425	4440	4456	2	3	5	6	8	9	11	13	1
28	4472	4487	4502	4518	4533	4548	4564	4579	4594	4609	2	3	5	6	8	9	11	12	1
29	4624	4639	4654	4669	4683	4698	4713	4728	4742	4757	1	3	4	6	7	9	10	12	1
30	4771	4786	4800	4814	4829	4843	4857	4871	4886	4900	1	3	4	6	7	9	10	11	1
31	4914	4928	4942	4955	4969	4983	4997	5011	5024	5038	1	3	4	6	7	8	10	11	1
32	5051	5065	5079	5092	5105	5119	5132	5145	5159	5172	1	3	4	5	7	8	9	11	1
33	5185	5198	5211	5224	5237	5250	5263	5276	5289	5302	1	3	4	5	6	8	9	10	1
34	5315	5328	5340	5353	5366	5378	5391	5403	5416	5428	1	3	4	5	6	8	9	10	1
35	5441	5453	5465	5478	5490	5502	5514	5527	5539	5551	1	2	4	5	6	7	9	10	1
36	5563	5575	5587	5599	5611	5623	5635	5647	5658	5670	1	2	4	5	6	7	8	10	1
37	5682	5694	5705	5717	5729	5740	5752	5763	5775	5786	1	2	3	5	6	7	8	9	1
38	5798	5809	5821	5832	5843	5855	5866	5877	5888	5899	1	2	3	5	6	7	8	9	1
39	5911	5922	5933	5944	5955	5966	5977	5988	5999	6010	1	2	3	4	5	7	8	9	1
40	6021	6031	6042	6053	6064	6075	6085	6096	6107	6117	1	2	3	4	5	6	8	9	1
41	6128	6138	6149	6160	6170	6180	6191	6201	6212	6222	1	2	3	4	5	6	7	8	
42	6232	6243	6253	6263	6274	6284	6294	6304	6314	6325	1	2	3	4	5	6	7	8	
43	6335	6345	6355	6365	6375	6385	6395	6405	6415	6425	1	2	3	4	5	6	7	8	
44	6435	6444	6454	6464	6474	6484	6493	6503	6513	6522	1	2	3	4	5	6	7	8	
45	6532	6542	6551	6561	6571	6580	6590	6599	6609	6618	1	2	3	4	5	6	7	8	
46	6628	6637	6646	6656	6665	6675	6684	6693	6702	6712	1	2	3	4	5	6	7	7	
47	6721	6730	6739	6749	6758	6767	6776	6785	6794	6803	1	2	3	4	5	5	6	7	
48	6812	6821	6830	6839	6848	6857	6866	6875	6884	6893	1	2	3	4	4	5	6	7	
49	6902	6911	6920	6928	6937	6946	6955	6964	6972	6981	1	2	3	4	4	5	6	7	
50	6990	6998	7007	7016	7024	7033	7042	7050	7059	7067	1	2	3	3	4	5	6	7	
51	7076	7084	7093	7101	7110	7118	7126	7135	7143	7152	1	2	3	3	4	5	6	7	
52	7160	7168	7177	7185	7193	7202	7210	7218	7226	7235	1	2	2	3	4	5	6	7	
53	7243	7251	7259	7267	7275	7284	7292	7300	7308	7316	1	2	2	3	4	5	6	6	
54	7324	7332	7340	7348	7356	7364	7372	7380	7388	7396	1	2	2	3	4	5	6	6	
	0	1	2	3	4	5	6	7	8	9	1	2	3	4	5	6	7	8	9

	0	1	2	3	4	5	6	7	8	9	Differences 1 2 3	4 5 6	7 8 9
55	7404	7412	7419	7427	7435	7443	7451	7459	7466	7474	1 2 2	3 4 5	5 6 7
56	7482	7490	7497	7505	7513	7520	7528	7536	7543	7551	1 2 2	3 4 5	5 6 7
57	7559	7566	7574	7582	7589	7597	7604	7612	7619	7627	1 2 2	3 4 5	5 6 7
58	7634	7642	7649	7657	7664	7672	7679	7686	7694	7701	1 1 2	3 4 4	5 6 7
59	7709	7716	7723	7731	7738	7745	7752	7760	7767	7774	1 1 2	3 4 4	5 6 7
60	7782	7789	7796	7803	7810	7818	7825	7832	7839	7846	1 1 2	3 4 4	5 6 6
61	7853	7860	7868	7875	7882	7889	7896	7903	7910	7917	1 1 2	3 4 4	5 6 6
62	7924	7931	7938	7945	7952	7959	7966	7973	7980	7987	1 1 2	3 3 4	5 6 6
63	7993	8000	8007	8014	8021	8028	8035	8041	8048	8055	1 1 2	3 3 4	5 5 6
64	8062	8069	8075	8082	8089	8096	8102	8109	8116	8122	1 1 2	3 3 4	5 5 6
65	8129	8136	8142	8149	8156	8162	8169	8176	8182	8189	1 1 2	3 3 4	5 5 6
66	8195	8202	8209	8215	8222	8228	8235	8241	8248	8254	1 1 2	3 3 4	5 5 6
67	8261	8267	8274	8280	8287	8293	8299	8306	8312	8319	1 1 2	3 3 4	5 5 6
68	8325	8331	8338	8344	8351	8357	8363	8370	8376	8382	1 1 2	3 3 4	4 5 6
69	8388	8395	8401	8407	8414	8420	8426	8432	8439	8445	1 1 2	2 3 4	4 5 6
70	8451	8457	8463	8470	8476	8482	8488	8494	8500	8506	1 1 2	2 3 4	4 5 6
71	8513	8519	8525	8531	8537	8543	8549	8555	8561	8567	1 1 2	2 3 4	4 5 5
72	8573	8579	8585	8591	8597	8603	8609	8615	8621	8627	1 1 2	2 3 4	4 5 5
73	8633	8639	8645	8651	8657	8663	8669	8675	8681	8686	1 1 2	2 3 4	4 5 5
74	8692	8698	8704	8710	8716	8722	8727	8733	8739	8745	1 1 2	2 3 4	4 5 5
75	8751	8756	8762	8768	8774	8779	8785	8791	8797	8802	1 1 2	2 3 3	4 5 5
76	8808	8814	8820	8825	8831	8837	8842	8848	8854	8859	1 1 2	2 3 3	4 5 5
77	8865	8871	8876	8882	8887	8893	8899	8904	8910	8915	1 1 2	2 3 3	4 4 5
78	8921	8927	8932	8938	8943	8949	8954	8960	8965	8971	1 1 2	2 3 3	4 4 5
79	8976	8982	8987	8993	8998	9004	9009	9015	9020	9025	1 1 2	2 3 3	4 4 5
80	9031	9036	9042	9047	9053	9058	9063	9069	9074	9079	1 1 2	2 3 3	4 4 5
81	9085	9090	9096	9101	9106	9112	9117	9122	9128	9133	1 1 2	2 3 3	4 4 5
82	9138	9143	9149	9154	9159	9165	9170	9175	9180	9186	1 1 2	2 3 3	4 4 5
83	9191	9196	9201	9206	9212	9217	9222	9227	9232	9238	1 1 2	2 3 3	4 4 5
84	9243	9248	9253	9258	9263	9269	9274	9279	9284	9289	1 1 2	2 3 3	4 4 5
85	9294	9299	9304	9309	9315	9320	9325	9330	9335	9340	1 1 2	2 3 3	4 4 5
86	9345	9350	9355	9360	9365	9370	9375	9380	9385	9390	1 1 2	2 3 3	4 4 5
87	9395	9400	9405	9410	9415	9420	9425	9430	9435	9440	0 1 1	2 2 3	3 4 4
88	9445	9450	9455	9460	9465	9469	9474	9479	9484	9489	0 1 1	2 2 3	3 4 4
89	9494	9499	9504	9509	9513	9518	9523	9528	9533	9538	0 1 1	2 2 3	3 4 4
90	9542	9547	9552	9557	9562	9566	9571	9576	9581	9586	0 1 1	2 2 3	3 4 4
91	9590	9595	9600	9605	9609	9614	9619	9624	9628	9633	0 1 1	2 2 3	3 4 4
92	9638	9643	9647	9652	9657	9661	9666	9671	9675	9680	0 1 1	2 2 3	3 4 4
93	9685	9689	9694	9699	9703	9708	9713	9717	9722	9727	0 1 1	2 2 3	3 4 4
94	9731	9736	9741	9745	9750	9754	9759	9763	9768	9773	0 1 1	2 2 3	3 4 4
95	9777	9782	9786	9791	9795	9800	9805	9809	9814	9818	0 1 1	2 2 3	3 4 4
96	9823	9827	9832	9836	9841	9845	9850	9854	9859	9863	0 1 1	2 2 3	3 4 4
97	9868	9872	9877	9881	9886	9890	9894	9899	9903	9908	0 1 1	2 2 3	3 4 4
98	9912	9917	9921	9926	9930	9934	9939	9943	9948	9952	0 1 1	2 2 3	3 4 4
99	9956	9961	9965	9969	9974	9978	9983	9987	9991	9996	0 1 1	2 2 3	3 3 4
	0	1	2	3	4	5	6	7	8	9	1 2 3	4 5 6	7 8 9

ANTILOGARITHMS

	0	1	2	3	4	5	6	7	8	9	Differences								
											1	2	3	4	5	6	7	8	9
·00	1000	1002	1005	1007	1009	1012	1014	1016	1019	1021	0	0	1	1	1	1	2	2	2
·01	1023	1026	1028	1030	1033	1035	1038	1040	1042	1045	0	0	1	1	1	1	2	2	2
·02	1047	1050	1052	1054	1057	1059	1062	1064	1067	1069	0	0	1	1	1	1	2	2	2
·03	1072	1074	1076	1079	1081	1084	1086	1089	1091	1094	0	0	1	1	1	1	2	2	2
·04	1096	1099	1102	1104	1107	1109	1112	1114	1117	1119	0	1	1	1	1	2	2	2	2
·05	1122	1125	1127	1130	1132	1135	1138	1140	1143	1146	0	1	1	1	1	2	2	2	2
·06	1148	1151	1153	1156	1159	1161	1164	1167	1169	1172	0	1	1	1	1	2	2	2	2
·07	1175	1178	1180	1183	1186	1189	1191	1194	1197	1199	0	1	1	1	1	2	2	2	2
·08	1202	1205	1208	1211	1213	1216	1219	1222	1225	1227	0	1	1	1	1	2	2	2	3
·09	1230	1233	1236	1239	1242	1245	1247	1250	1253	1256	0	1	1	1	1	2	2	2	3
·10	1259	1262	1265	1268	1271	1274	1276	1279	1282	1285	0	1	1	1	1	2	2	2	3
·11	1288	1291	1294	1297	1300	1303	1306	1309	1312	1315	0	1	1	1	2	2	2	2	3
·12	1318	1321	1324	1327	1330	1334	1337	1340	1343	1346	0	1	1	1	2	2	2	3	3
·13	1349	1352	1355	1358	1361	1365	1368	1371	1374	1377	0	1	1	1	2	2	2	3	3
·14	1380	1384	1387	1390	1393	1396	1400	1403	1406	1409	0	1	1	1	2	2	2	3	3
·15	1413	1416	1419	1422	1426	1429	1432	1435	1439	1442	0	1	1	1	2	2	2	3	3
·16	1445	1449	1452	1455	1459	1462	1466	1469	1472	1476	0	1	1	1	2	2	2	3	3
·17	1479	1483	1486	1489	1493	1496	1500	1503	1507	1510	0	1	1	1	2	2	2	3	3
·18	1514	1517	1521	1524	1528	1531	1535	1538	1542	1545	0	1	1	1	2	2	2	3	3
·19	1549	1552	1556	1560	1563	1567	1570	1574	1578	1581	0	1	1	1	2	2	3	3	3
·20	1585	1589	1592	1596	1600	1603	1607	1611	1614	1618	0	1	1	1	2	2	3	3	3
·21	1622	1626	1629	1633	1637	1641	1644	1648	1652	1656	0	1	1	2	2	2	3	3	3
·22	1660	1663	1667	1671	1675	1679	1683	1687	1690	1694	0	1	1	2	2	2	3	3	3
·23	1698	1702	1706	1710	1714	1718	1722	1726	1730	1734	0	1	1	2	2	2	3	3	4
·24	1738	1742	1746	1750	1754	1758	1762	1766	1770	1774	0	1	1	2	2	2	3	3	4
·25	1778	1782	1786	1791	1795	1799	1803	1807	1811	1816	0	1	1	2	2	2	3	3	4
·26	1820	1824	1828	1832	1837	1841	1845	1849	1854	1858	0	1	1	2	2	3	3	3	4
·27	1862	1866	1871	1875	1879	1884	1888	1892	1897	1901	0	1	1	2	2	3	3	3	4
·28	1905	1910	1914	1919	1923	1928	1932	1936	1941	1945	0	1	1	2	2	3	3	4	4
·29	1950	1954	1959	1963	1968	1972	1977	1982	1986	1991	0	1	1	2	2	3	3	4	4
·30	1995	2000	2004	2009	2014	2018	2023	2028	2032	2037	0	1	1	2	2	3	3	4	4
·31	2042	2046	2051	2056	2061	2065	2070	2075	2080	2084	0	1	1	2	2	3	3	4	4
·32	2089	2094	2099	2104	2109	2113	2118	2123	2128	2133	0	1	1	2	2	3	3	4	4
·33	2138	2143	2148	2153	2158	2163	2168	2173	2178	2183	0	1	1	2	2	3	3	4	4
·34	2188	2193	2198	2203	2208	2213	2218	2223	2228	2234	1	1	2	2	3	3	4	4	5
·35	2239	2244	2249	2254	2259	2265	2270	2275	2280	2286	1	1	2	2	3	3	4	4	5
·36	2291	2296	2301	2307	2312	2317	2323	2328	2333	2339	1	1	2	2	3	3	4	4	5
·37	2344	2350	2355	2360	2366	2371	2377	2382	2388	2393	1	1	2	2	3	3	4	4	5
·38	2399	2404	2410	2415	2421	2427	2432	2438	2443	2449	1	1	2	2	3	3	4	4	5
·39	2455	2460	2466	2472	2477	2483	2489	2495	2500	2506	1	1	2	2	3	3	4	5	5
·40	2512	2518	2523	2529	2535	2541	2547	2553	2559	2564	1	1	2	2	3	4	4	5	5
·41	2570	2576	2582	2588	2594	2600	2606	2612	2618	2624	1	1	2	2	3	4	4	5	5
·42	2630	2636	2642	2649	2655	2661	2667	2673	2679	2685	1	1	2	2	3	4	4	5	6
·43	2692	2698	2704	2710	2716	2723	2729	2735	2742	2748	1	1	2	3	3	4	4	5	6
·44	2754	2761	2767	2773	2780	2786	2793	2799	2805	2812	1	1	2	3	3	4	4	5	6
·45	2818	2825	2831	2838	2844	2851	2858	2864	2871	2877	1	1	2	3	3	4	5	5	6
·46	2884	2891	2897	2904	2911	2917	2924	2931	2938	2944	1	1	2	3	3	4	5	5	6
·47	2951	2958	2965	2972	2979	2985	2992	2999	3006	3013	1	1	2	3	3	4	5	5	6
·48	3020	3027	3034	3041	3048	3055	3062	3069	3076	3083	1	1	2	3	4	4	5	6	6
·49	3090	3097	3105	3112	3119	3126	3133	3141	3148	3155	1	1	2	3	4	4	5	6	6
	0	1	2	3	4	5	6	7	8	9	1	2	3	4	5	6	7	8	9

	0	1	2	3	4	5	6	7	8	9	1	2	3	4	5	6	7	8	9
															Differences				
50	3162	3170	3177	3184	3192	3199	3206	3214	3221	3228	1	1	2	3	4	4	5	6	7
51	3236	3243	3251	3258	3266	3273	3281	3289	3296	3304	1	2	2	3	4	5	5	6	7
52	3311	3319	3327	3334	3342	3350	3357	3365	3373	3381	1	2	2	3	4	5	5	6	7
53	3388	3396	3404	3412	3420	3428	3436	3443	3451	3459	1	2	2	3	4	5	6	6	7
54	3467	3475	3483	3491	3499	3508	3516	3524	3532	3540	1	2	2	3	4	5	6	6	7
55	3548	3556	3565	3573	3581	3589	3597	3606	3614	3622	1	2	2	3	4	5	6	7	7
56	3631	3639	3648	3656	3664	3673	3681	3690	3698	3707	1	2	3	3	4	5	6	7	8
57	3715	3724	3733	3741	3750	3758	3767	3776	3784	3793	1	2	3	3	4	5	6	7	8
58	3802	3811	3819	3828	3837	3846	3855	3864	3873	3882	1	2	3	4	4	5	6	7	8
59	3890	3899	3908	3917	3926	3936	3945	3954	3963	3972	1	2	3	4	5	5	6	7	8
60	3981	3990	3999	4009	4018	4027	4036	4046	4055	4064	1	2	3	4	5	6	6	7	8
61	4074	4083	4093	4102	4111	4121	4130	4140	4150	4159	1	2	3	4	5	6	7	8	9
62	4169	4178	4188	4198	4207	4217	4227	4236	4246	4256	1	2	3	4	5	6	7	8	9
63	4266	4276	4285	4295	4305	4315	4325	4335	4345	4355	1	2	3	4	5	6	7	8	9
64	4365	4375	4385	4395	4406	4416	4426	4436	4446	4457	1	2	3	4	5	6	7	8	9
65	4467	4477	4487	4498	4508	4519	4529	4539	4550	4560	1	2	3	4	5	6	7	8	9
66	4571	4581	4592	4603	4613	4624	4634	4645	4656	4667	1	2	3	4	5	6	7	8	10
67	4677	4688	4699	4710	4721	4732	4742	4753	4764	4775	1	2	3	4	5	7	8	9	10
68	4786	4797	4808	4819	4831	4842	4853	4864	4875	4887	1	2	3	4	6	7	8	9	10
69	4898	4909	4920	4932	4943	4955	4966	4977	4989	5000	1	2	3	5	6	7	8	9	10
70	5012	5023	5035	5047	5058	5070	5082	5093	5105	5117	1	2	4	5	6	7	8	9	11
71	5129	5140	5152	5164	5176	5188	5200	5212	5224	5236	1	2	4	5	6	7	8	10	11
72	5248	5260	5272	5284	5297	5309	5321	5333	5346	5358	1	2	4	5	6	7	9	10	11
73	5370	5383	5395	5408	5420	5433	5445	5458	5470	5483	1	3	4	5	6	8	9	10	11
74	5495	5508	5521	5534	5546	5559	5572	5585	5598	5610	1	3	4	5	6	8	9	10	12
75	5623	5636	5649	5662	5675	5689	5702	5715	5728	5741	1	3	4	5	7	8	9	10	12
76	5754	5768	5781	5794	5808	5821	5834	5848	5861	5875	1	3	4	5	7	8	9	11	12
77	5888	5902	5916	5929	5943	5957	5970	5984	5998	6012	1	3	4	5	7	8	10	11	12
78	6026	6039	6053	6067	6081	6095	6109	6124	6138	6152	1	3	4	6	7	8	10	11	13
79	6166	6180	6194	6209	6223	6237	6252	6266	6281	6295	1	3	4	6	7	9	10	11	13
80	6310	6324	6339	6353	6368	6383	6397	6412	6427	6442	1	3	4	6	7	9	10	12	13
81	6457	6471	6486	6501	6516	6531	6546	6561	6577	6592	2	3	5	6	8	9	11	12	14
82	6607	6622	6637	6653	6668	6683	6699	6714	6730	6745	2	3	5	6	8	9	11	12	14
83	6761	6776	6792	6808	6823	6839	6855	6871	6887	6902	2	3	5	6	8	9	11	13	14
84	6918	6934	6950	6966	6982	6998	7015	7031	7047	7063	2	3	5	6	8	10	11	13	15
85	7079	7096	7112	7129	7145	7161	7178	7194	7211	7228	2	3	5	7	8	10	12	13	15
86	7244	7261	7278	7295	7311	7328	7345	7362	7379	7396	2	3	5	7	8	10	12	13	15
87	7413	7430	7447	7464	7482	7499	7516	7534	7551	7568	2	3	5	7	9	10	12	14	16
88	7586	7603	7621	7638	7656	7674	7691	7709	7727	7745	2	4	5	7	9	11	12	14	16
89	7762	7780	7798	7816	7834	7852	7870	7889	7907	7925	2	4	5	7	9	11	13	14	16
90	7943	7962	7980	7998	8017	8035	8054	8072	8091	8110	2	4	6	7	9	11	13	15	17
91	8128	8147	8166	8185	8204	8222	8241	8260	8279	8299	2	4	6	7	9	11	13	15	17
92	8318	8337	8356	8375	8395	8414	8433	8453	8472	8492	2	4	6	8	10	12	14	15	17
93	8511	8531	8551	8570	8590	8610	8630	8650	8670	8690	2	4	6	8	10	12	14	16	18
94	8710	8730	8750	8770	8790	8810	8831	8851	8872	8892	2	4	6	8	10	12	14	16	18
95	8913	8933	8954	8974	8995	9016	9036	9057	9078	9099	2	4	6	8	10	12	15	17	19
96	9120	9141	9162	9183	9204	9226	9247	9268	9290	9311	2	4	6	8	11	13	15	17	19
97	9333	9354	9376	9397	9419	9441	9462	9484	9506	9528	2	4	7	9	11	13	15	17	20
98	9550	9572	9594	9616	9638	9661	9683	9705	9727	9750	2	4	7	9	11	13	16	18	20
99	9772	9795	9817	9840	9863	9886	9908	9931	9954	9977	2	5	7	9	11	14	16	18	20
	0	1	2	3	4	5	6	7	8	9	1	2	3	4	5	6	7	8	9

SQUARES OF NUMBERS

	0	1	2	3	4	5	6	7	8	9	Differences							
											1	2	3	4	5	6	7	8
10	1000	1020	1040	1061	1082	1103	1124	1145	1166	1188	2	4	6	8	10	13	15	17 1
11	1210	1232	1254	1277	1300	1323	1346	1369	1392	1416	2	5	7	9	11	14	16	18 2
12	1440	1464	1488	1513	1538	1563	1588	1613	1638	1664	2	5	7	10	12	15	17	20 2
13	1690	1716	1742	1769	1796	1823	1850	1877	1904	1932	3	5	8	11	13	16	19	22 2
14	1960	1988	2016	2045	2074	2103	2132	2161	2190	2220	3	6	9	12	14	17	20	23 2
15	2250	2280	2310	2341	2372	2403	2434	2465	2496	2528	3	6	9	12	15	19	22	25 2
16	2560	2592	2624	2657	2690	2723	2756	2789	2822	2856	3	7	10	13	16	20	23	26 3
17	2890	2924	2958	2993	3028	3063	3098	3133	3168	3204	3	7	10	14	17	21	24	28 3
18	3240	3276	3312	3349	3386	3423	3460	3497	3534	3572	4	7	11	15	18	22	26	30 3
19	3610	3648	3686	3725	3764	3803	3842	3881	3920	3960	4	8	12	16	19	23	27	31 3
20	4000	4040	4080	4121	4162	4203	4244	4285	4326	4368	4	8	12	16	20	25	29	33 3
21	4410	4452	4494	4537	4580	4623	4666	4709	4752	4796	4	9	13	17	21	26	30	34 3
22	4840	4884	4928	4973	5018	5063	5108	5153	5198	5244	4	9	13	18	22	27	31	36 4
23	5290	5336	5382	5429	5476	5523	5570	5617	5664	5712	5	9	14	19	23	28	33	38 4
24	5760	5808	5856	5905	5954	6003	6052	6101	6150	6200	5	10	15	20	24	29	34	39 4
25	6250	6300	6350	6401	6452	6503	6554	6605	6656	6708	5	10	15	20	25	31	36	41 4
26	6760	6812	6864	6917	6970	7023	7076	7129	7182	7236	5	11	16	21	26	32	37	42 4
27	7290	7344	7398	7453	7508	7563	7618	7673	7728	7784	5	11	16	22	27	33	38	44 4
28	7840	7896	7952	8009	8066	8123	8180	8237	8294	8352	6	11	17	23	28	34	40	46 5
29	8410	8468	8526	8585	8644	8703	8762	8821	8880	8940	6	12	18	24	29	35	41	47 5
30	9000	9060	9120	9181	9242	9303	9364	9425	9486	9548	6	12	18	24	30	37	43	49 5
31	9610	9672	9734	9797	9860	9923	9986	—	—	—	6	13	19	25	31	38	44	50 5
31	—	—	—	—	—	—	—	1005	1011	1018	1	1	2	3	3	4	5	5
32	1024	1030	1037	1043	1050	1056	1063	1069	1076	1082	1	1	2	3	3	4	5	5
33	1089	1096	1102	1109	1116	1122	1129	1136	1142	1149	1	1	2	3	3	4	5	5
34	1156	1163	1170	1176	1183	1190	1197	1204	1211	1218	1	1	2	3	3	4	5	6
35	1225	1232	1239	1246	1253	1260	1267	1274	1282	1289	1	1	2	3	4	4	5	6
36	1296	1303	1310	1318	1325	1332	1340	1347	1354	1362	1	1	2	3	4	4	5	6
37	1369	1376	1384	1391	1399	1406	1414	1421	1429	1436	1	2	2	3	4	5	5	6
38	1444	1452	1459	1467	1475	1482	1490	1498	1505	1513	1	2	2	3	4	5	5	6
39	1521	1529	1537	1544	1552	1560	1568	1576	1584	1592	1	2	2	3	4	5	6	6
40	1600	1608	1616	1624	1632	1640	1648	1656	1665	1673	1	2	2	3	4	5	6	6
41	1681	1689	1697	1706	1714	1722	1731	1739	1747	1756	1	2	2	3	4	5	6	7
42	1764	1772	1781	1789	1798	1806	1815	1823	1832	1840	1	2	3	3	4	5	6	7
43	1849	1858	1866	1875	1884	1892	1901	1910	1918	1927	1	2	3	3	4	5	6	7
44	1936	1945	1954	1962	1971	1980	1989	1998	2007	2016	1	2	3	4	5	5	6	7
45	2025	2034	2043	2052	2061	2070	2079	2088	2098	2107	1	2	3	4	5	5	6	7
46	2116	2125	2134	2144	2153	2162	2172	2181	2190	2200	1	2	3	4	5	6	7	7
47	2209	2218	2228	2237	2247	2256	2266	2275	2285	2294	1	2	3	4	5	6	7	8
48	2304	2314	2323	2333	2343	2352	2362	2372	2381	2391	1	2	3	4	5	6	7	8
49	2401	2411	2421	2430	2440	2450	2460	2470	2480	2490	1	2	3	4	5	6	7	8
50	2500	2510	2520	2530	2540	2550	2560	2570	2581	2591	1	2	3	4	5	6	7	8
51	2601	2611	2621	2632	2642	2652	2663	2673	2683	2694	1	2	3	4	5	6	7	8
52	2704	2714	2725	2735	2746	2756	2767	2777	2788	2798	1	2	3	4	5	6	7	8
53	2809	2820	2830	2841	2852	2862	2873	2884	2894	2905	1	2	3	4	5	6	7	9 1
54	2916	2927	2938	2948	2959	2970	2981	2992	3003	3014	1	2	3	4	5	7	8	9 1

	0	1	2	3	4	5	6	7	8	9	Differences								
											1	2	3	4	5	6	7	8	9
55	3025	3036	3047	3058	3069	3080	3091	3102	3114	3125	1	2	3	4	6	7	8	9	10
56	3136	3147	3158	3170	3181	3192	3204	3215	3226	3238	1	2	3	5	6	7	8	9	10
57	3249	3260	3272	3283	3295	3306	3318	3329	3341	3352	1	2	3	5	6	7	8	9	10
58	3364	3376	3387	3399	3411	3422	3434	3446	3457	3469	1	2	4	5	6	7	8	9	11
59	3481	3493	3505	3516	3528	3540	3552	3564	3576	3588	1	2	4	5	6	7	8	10	11
60	3600	3612	3624	3636	3648	3660	3672	3684	3697	3709	1	2	4	5	6	7	8	10	11
61	3721	3733	3745	3758	3770	3782	3795	3807	3819	3832	1	2	4	5	6	7	9	10	11
62	3844	3856	3869	3881	3894	3906	3919	3931	3944	3956	1	2	4	5	6	7	9	10	11
63	3969	3982	3994	4007	4020	4032	4045	4058	4070	4083	1	3	4	5	6	8	9	10	11
64	4096	4109	4122	4134	4147	4160	4173	4186	4199	4212	1	3	4	5	6	8	9	10	12
65	4225	4238	4251	4264	4277	4290	4303	4316	4330	4343	1	3	4	5	7	8	9	10	12
66	4356	4369	4382	4396	4409	4422	4436	4449	4462	4476	1	3	4	5	7	8	9	11	12
67	4489	4502	4516	4529	4543	4556	4570	4583	4597	4610	1	3	4	5	7	8	9	11	12
68	4624	4638	4651	4665	4679	4692	4706	4720	4733	4747	1	3	4	5	7	8	10	11	12
69	4761	4775	4789	4802	4816	4830	4844	4858	4872	4886	1	3	4	6	7	8	10	11	13
70	4900	4914	4928	4942	4956	4970	4984	4998	5013	5027	1	3	4	6	7	8	10	11	13
71	5041	5055	5069	5084	5098	5112	5127	5141	5155	5170	1	3	4	6	7	9	10	11	13
72	5184	5198	5213	5227	5242	5256	5271	5285	5300	5314	1	3	4	6	7	9	10	11	13
73	5329	5344	5358	5373	5388	5402	5417	5432	5446	5461	1	3	4	6	7	9	10	12	13
74	5476	5491	5506	5520	5535	5550	5565	5580	5595	5610	1	3	4	6	7	9	10	12	13
75	5625	5640	5655	5670	5685	5700	5715	5730	5746	5761	2	3	5	6	8	9	11	12	14
76	5776	5791	5806	5822	5837	5852	5868	5883	5898	5914	2	3	5	6	8	9	11	12	14
77	5929	5944	5960	5975	5991	6006	6022	6037	6053	6068	2	3	5	6	8	9	11	12	14
78	6084	6100	6115	6131	6147	6162	6178	6194	6209	6225	2	3	5	6	8	9	11	13	14
79	6241	6257	6273	6288	6304	6320	6336	6352	6368	6384	2	3	5	6	8	10	11	13	14
80	6400	6416	6432	6448	6464	6480	6496	6512	6529	6545	2	3	5	6	8	10	11	13	14
81	6561	6577	6593	6610	6626	6642	6659	6675	6691	6708	2	3	5	7	8	10	11	13	15
82	6724	6740	6757	6773	6790	6806	6823	6839	6856	6872	2	3	5	7	8	10	12	13	15
83	6889	6906	6922	6939	6956	6972	6989	7006	7022	7039	2	3	5	7	8	10	12	13	15
84	7056	7073	7090	7106	7123	7140	7157	7174	7191	7208	2	3	5	7	8	10	12	14	15
85	7225	7242	7259	7276	7293	7310	7327	7344	7362	7379	2	3	5	7	9	10	12	14	15
86	7396	7413	7430	7448	7465	7482	7500	7517	7534	7552	2	3	5	7	9	10	12	14	16
87	7569	7586	7604	7621	7639	7656	7674	7691	7709	7726	2	3	5	7	9	10	12	14	16
88	7744	7762	7779	7797	7815	7832	7850	7868	7885	7903	2	4	5	7	9	11	12	14	16
89	7921	7939	7957	7974	7992	8010	8028	8046	8064	8082	2	4	5	7	9	11	13	14	16
90	8100	8118	8136	8154	8172	8190	8208	8226	8245	8263	2	4	5	7	9	11	13	14	16
91	8281	8299	8317	8336	8354	8372	8391	8409	8427	8446	2	4	5	7	9	11	13	15	16
92	8464	8482	8501	8519	8538	8556	8575	8593	8612	8630	2	4	6	7	9	11	13	15	17
93	8649	8668	8686	8705	8724	8742	8761	8780	8798	8817	2	4	6	7	9	11	13	15	17
94	8836	8855	8874	8892	8911	8930	8949	8968	8987	9006	2	4	6	8	9	11	13	15	17
95	9025	9044	9063	9082	9101	9120	9139	9158	9178	9197	2	4	6	8	10	11	13	15	17
96	9216	9235	9254	9274	9293	9312	9332	9351	9370	9390	2	4	6	8	10	12	14	15	17
97	9409	9428	9448	9467	9487	9506	9526	9545	9565	9584	2	4	6	8	10	12	14	16	18
98	9604	9624	9643	9663	9683	9702	9722	9742	9761	9781	2	4	6	8	10	12	14	16	18
99	9801	9821	9841	9860	9880	9900	9920	9940	9960	9980	2	4	6	8	10	12	14	16	18

SQUARE ROOTS OF NUMBERS 1–10

	0	1	2	3	4	5	6	7	8	9	Differences 1 2 3	4 5 6	7 8
1·0	1·000	1·005	1·010	1·015	1·020	1·025	1·030	1·034	1·039	1·044	0 1 1	2 2 3	3 4
1·1	1·049	1·054	1·058	1·063	1·068	1·072	1·077	1·082	1·086	1·091	0 1 1	2 2 3	3 4
1·2	1·095	1·100	1·105	1·109	1·114	1·118	1·122	1·127	1·131	1·136	0 1 1	2 2 3	3 4
1·3	1·140	1·145	1·149	1·153	1·158	1·162	1·166	1·170	1·175	1·179	0 1 1	2 2 3	3 3
1·4	1·183	1·187	1·192	1·196	1·200	1·204	1·208	1·212	1·217	1·221	0 1 1	2 2 2	3 3
1·5	1·225	1·229	1·233	1·237	1·241	1·245	1·249	1·253	1·257	1·261	0 1 1	2 2 2	3 3
1·6	1·265	1·269	1·273	1·277	1·281	1·285	1·288	1·292	1·296	1·300	0 1 1	2 2 2	3 3
1·7	1·304	1·308	1·311	1·315	1·319	1·323	1·327	1·330	1·334	1·338	0 1 1	2 2 2	3 3
1·8	1·342	1·345	1·349	1·353	1·356	1·360	1·364	1·367	1·371	1·375	0 1 1	1 2 2	3 3
1·9	1·378	1·382	1·386	1·389	1·393	1·396	1·400	1·404	1·407	1·411	0 1 1	1 2 2	3 3
2·0	1·414	1·418	1·421	1·425	1·428	1·432	1·435	1·439	1·442	1·446	0 1 1	1 2 2	2 3
2·1	1·449	1·453	1·456	1·459	1·463	1·466	1·470	1·473	1·476	1·480	0 1 1	1 2 2	2 3
2·2	1·483	1·487	1·490	1·493	1·497	1·500	1·503	1·507	1·510	1·513	0 1 1	1 2 2	2 3
2·3	1·517	1·520	1·523	1·526	1·530	1·533	1·536	1·539	1·543	1·546	0 1 1	1 2 2	2 3
2·4	1·549	1·552	1·556	1·559	1·562	1·565	1·568	1·572	1·575	1·578	0 1 1	1 2 2	2 3
2·5	1·581	1·584	1·587	1·591	1·594	1·597	1·600	1·603	1·606	1·609	0 1 1	1 2 2	2 3
2·6	1·612	1·616	1·619	1·622	1·625	1·628	1·631	1·634	1·637	1·640	0 1 1	1 2 2	2 2
2·7	1·643	1·646	1·649	1·652	1·655	1·658	1·661	1·664	1·667	1·670	0 1 1	1 2 2	2 2
2·8	1·673	1·676	1·679	1·682	1·685	1·688	1·691	1·694	1·697	1·700	0 1 1	1 1 2	2 2
2·9	1·703	1·706	1·709	1·712	1·715	1·718	1·720	1·723	1·726	1·729	0 1 1	1 1 2	2 2
3·0	1·732	1·735	1·738	1·741	1·744	1·746	1·749	1·752	1·755	1·758	0 1 1	1 1 2	2 2
3·1	1·761	1·764	1·766	1·769	1·772	1·775	1·778	1·780	1·783	1·786	0 1 1	1 1 2	2 2
3·2	1·789	1·792	1·794	1·797	1·800	1·803	1·806	1·808	1·811	1·814	0 1 1	1 1 2	2 2
3·3	1·817	1·819	1·822	1·825	1·828	1·830	1·833	1·836	1·838	1·841	0 1 1	1 1 2	2 2
3·4	1·844	1·847	1·849	1·852	1·855	1·857	1·860	1·863	1·865	1·868	0 1 1	1 1 2	2 2
3·5	1·871	1·873	1·876	1·879	1·881	1·884	1·887	1·889	1·892	1·895	0 1 1	1 1 2	2 2
3·6	1·897	1·900	1·903	1·905	1·908	1·910	1·913	1·916	1·918	1·921	0 1 1	1 1 2	2 2
3·7	1·924	1·926	1·929	1·931	1·934	1·936	1·939	1·942	1·944	1·947	0 1 1	1 1 2	2 2
3·8	1·949	1·952	1·954	1·957	1·960	1·962	1·965	1·967	1·970	1·972	0 1 1	1 1 2	2 2
3·9	1·975	1·977	1·980	1·982	1·985	1·987	1·990	1·992	1·995	1·997	0 1 1	1 1 2	2 2
4·0	2·000	2·002	2·005	2·007	2·010	2·012	2·015	2·017	2·020	2·022	0 0 1	1 1 1	2 2
4·1	2·025	2·027	2·030	2·032	2·035	2·037	2·040	2·042	2·045	2·047	0 0 1	1 1 1	2 2
4·2	2·049	2·052	2·054	2·057	2·059	2·062	2·064	2·066	2·069	2·071	0 0 1	1 1 1	2 2
4·3	2·074	2·076	2·078	2·081	2·083	2·086	2·088	2·090	2·093	2·095	0 0 1	1 1 1	2 2
4·4	2·098	2·100	2·102	2·105	2·107	2·110	2·112	2·114	2·117	2·119	0 0 1	1 1 1	2 2
4·5	2·121	2·124	2·126	2·128	2·131	2·133	2·135	2·138	2·140	2·142	0 0 1	1 1 1	2 2
4·6	2·145	2·147	2·149	2·152	2·154	2·156	2·159	2·161	2·163	2·166	0 0 1	1 1 1	2 2
4·7	2·168	2·170	2·173	2·175	2·177	2·179	2·182	2·184	2·186	2·189	0 0 1	1 1 1	2 2
4·8	2·191	2·193	2·195	2·198	2·200	2·202	2·205	2·207	2·209	2·211	0 0 1	1 1 1	2 2
4·9	2·214	2·216	2·218	2·220	2·223	2·225	2·227	2·229	2·232	2·234	0 0 1	1 1 1	2 2
5·0	2·236	2·238	2·241	2·243	2·245	2·247	2·249	2·252	2·254	2·256	0 0 1	1 1 1	2 2
5·1	2·258	2·261	2·263	2·265	2·267	2·269	2·272	2·274	2·276	2·278	0 0 1	1 1 1	2 2
5·2	2·280	2·283	2·285	2·287	2·289	2·291	2·293	2·296	2·298	2·300	0 0 1	1 1 1	2 2
5·3	2·302	2·304	2·307	2·309	2·311	2·313	2·315	2·317	2·319	2·322	0 0 1	1 1 1	2 2
5·4	2·324	2·326	2·328	2·330	2·332	2·335	2·337	2·339	2·341	2·343	0 0 1	1 1 1	1 2

	0	1	2	3	4	5	6	7	8	9	Differences 1 2 3	4 5 6	7 8 9
·5	2·345	2·347	2·349	2·352	2·354	2·356	2·358	2·360	2·362	2·364	0 0 1	1 1 1	1 2 2
·6	2·366	2·369	2·371	2·373	2·375	2·377	2·379	2·381	2·383	2·385	0 0 1	1 1 1	1 2 2
·7	2·387	2·390	2·392	2·394	2·396	2·398	2·400	2·402	2·404	2·406	0 0 1	1 1 1	1 2 2
·8	2·408	2·410	2·412	2·415	2·417	2·419	2·421	2·423	2·425	2·427	0 0 1	1 1 1	1 2 2
·9	2·429	2·431	2·433	2·435	2·437	2·439	2·441	2·443	2·445	2·447	0 0 1	1 1 1	1 2 2
·0	2·449	2·452	2·454	2·456	2·458	2·460	2·462	2·464	2·466	2·468	0 0 1	1 1 1	1 2 2
·1	2·470	2·472	2·474	2·476	2·478	2·480	2·482	2·484	2·486	2·488	0 0 1	1 1 1	1 2 2
·2	2·490	2·492	2·494	2·496	2·498	2·500	2·502	2·504	2·506	2·508	0 0 1	1 1 1	1 2 2
·3	2·510	2·512	2·514	2·516	2·518	2·520	2·522	2·524	2·526	2·528	0 0 1	1 1 1	1 2 2
·4	2·530	2·532	2·534	2·536	2·538	2·540	2·542	2·544	2·546	2·548	0 0 1	1 1 1	1 2 2
·5	2·550	2·551	2·553	2·555	2·557	2·559	2·561	2·563	2·565	2·567	0 0 1	1 1 1	1 2 2
·6	2·569	2·571	2·573	2·575	2·577	2·579	2·581	2·583	2·585	2·587	0 0 1	1 1 1	1 2 2
·7	2·588	2·590	2·592	2·594	2·596	2·598	2·600	2·602	2·604	2·606	0 0 1	1 1 1	1 2 2
·8	2·608	2·610	2·612	2·613	2·615	2·617	2·619	2·621	2·623	2·625	0 0 1	1 1 1	1 2 2
·9	2·627	2·629	2·631	2·632	2·634	2·636	2·638	2·640	2·642	2·644	0 0 1	1 1 1	1 2 2
·0	2·646	2·648	2·650	2·651	2·653	2·655	2·657	2·659	2·661	2·663	0 0 1	1 1 1	1 2 2
·1	2·665	2·666	2·668	2·670	2·672	2·674	2·676	2·678	2·680	2·681	0 0 1	1 1 1	1 1 2
·2	2·683	2·685	2·687	2·689	2·691	2·693	2·694	2·696	2·698	2·700	0 0 1	1 1 1	1 1 2
·3	2·702	2·704	2·706	2·707	2·709	2·711	2·713	2·715	2·717	2·718	0 0 1	1 1 1	1 1 2
·4	2·720	2·722	2·724	2·726	2·728	2·729	2·731	2·733	2·735	2·737	0 0 1	1 1 1	1 1 2
·5	2·739	2·740	2·742	2·744	2·746	2·748	2·750	2·751	2·753	2·755	0 0 1	1 1 1	1 1 2
·6	2·757	2·759	2·760	2·762	2·764	2·766	2·768	2·769	2·771	2·773	0 0 1	1 1 1	1 1 2
·7	2·775	2·777	2·778	2·780	2·782	2·784	2·786	2·787	2·789	2·791	0 0 1	1 1 1	1 1 2
·8	2·793	2·795	2·796	2·798	2·800	2·802	2·804	2·805	2·807	2·809	0 0 1	1 1 1	1 1 2
·9	2·811	2·812	2·814	2·816	2·818	2·820	2·821	2·823	2·825	2·827	0 0 1	1 1 1	1 1 2
·0	2·828	2·830	2·832	2·834	2·835	2·837	2·839	2·841	2·843	2·844	0 0 1	1 1 1	1 1 2
·1	2·846	2·848	2·850	2·851	2·853	2·855	2·857	2·858	2·860	2·862	0 0 1	1 1 1	1 1 2
·2	2·864	2·865	2·867	2·869	2·871	2·872	2·874	2·876	2·877	2·879	0 0 1	1 1 1	1 1 2
·3	2·881	2·883	2·884	2·886	2·888	2·890	2·891	2·893	2·895	2·897	0 0 1	1 1 1	1 1 2
·4	2·898	2·900	2·902	2·903	2·905	2·907	2·909	2·910	2·912	2·914	0 0 1	1 1 1	1 1 2
·5	2·915	2·917	2·919	2·921	2·922	2·924	2·926	2·927	2·929	2·931	0 0 1	1 1 1	1 1 2
·6	2·933	2·934	2·936	2·938	2·939	2·941	2·943	2·944	2·946	2·948	0 0 1	1 1 1	1 1 2
·7	2·950	2·951	2·953	2·955	2·956	2·958	2·960	2·961	2·963	2·965	0 0 1	1 1 1	1 1 2
·8	2·966	2·968	2·970	2·972	2·973	2·975	2·977	2·978	2·980	2·982	0 0 1	1 1 1	1 1 2
·9	2·983	2·985	2·987	2·988	2·990	2·992	2·993	2·995	2·997	2·998	0 0 1	1 1 1	1 1 2
·0	3·000	3·002	3·003	3·005	3·007	3·008	3·010	3·012	3·013	3·015	0 0 0	1 1 1	1 1 1
·1	3·017	3·018	3·020	3·022	3·023	3·025	3·027	3·028	3·030	3·032	0 0 0	1 1 1	1 1 1
·2	3·033	3·035	3·036	3·038	3·040	3·041	3·043	3·045	3·046	3·048	0 0 0	1 1 1	1 1 1
·3	3·050	3·051	3·053	3·055	3·056	3·058	3·059	3·061	3·063	3·064	0 0 0	1 1 1	1 1 1
·4	3·066	3·068	3·069	3·071	3·072	3·074	3·076	3·077	3·079	3·081	0 0 0	1 1 1	1 1 1
·5	3·082	3·084	3·085	3·087	3·089	3·090	3·092	3·094	3·095	3·097	0 0 0	1 1 1	1 1 1
·6	3·098	3·100	3·102	3·103	3·105	3·106	3·108	3·110	3·111	3·113	0 0 0	1 1 1	1 1 1
·7	3·114	3·116	3·118	3·119	3·121	3·122	3·124	3·126	3·127	3·129	0 0 0	1 1 1	1 1 1
·8	3·130	3·132	3·134	3·135	3·137	3·138	3·140	3·142	3·143	3·145	0 0 0	1 1 1	1 1 1
·9	3·146	3·148	3·150	3·151	3·153	3·154	3·156	3·158	3·159	3·161	0 0 0	1 1 1	1 1 1

	0	1	2	3	4	5	6	7	8	9	Differences 1 2 3	4 5 6	7 8
10	3·162	3·178	3·194	3·209	3·225	3·240	3·256	3·271	3·286	3·302	2 3 5	6 8 9	11 12
11	3·317	3·332	3·347	3·362	3·376	3·391	3·406	3·421	3·435	3·450	1 3 4	6 7 9	10 12
12	3·464	3·479	3·493	3·507	3·521	3·536	3·550	3·564	3·578	3·592	1 3 4	6 7 8	10 11
13	3·606	3·619	3·633	3·647	3·661	3·674	3·688	3·701	3·715	3·728	1 3 4	5 7 8	10 11
14	3·742	3·755	3·768	3·782	3·795	3·808	3·821	3·834	3·847	3·860	1 3 4	5 7 8	9 11
15	3·873	3·886	3·899	3·912	3·924	3·937	3·950	3·962	3·975	3·987	1 3 4	5 6 8	9 10
16	4·000	4·012	4·025	4·037	4·050	4·062	4·074	4·087	4·099	4·111	1 2 4	5 6 7	9 10
17	4·123	4·135	4·147	4·159	4·171	4·183	4·195	4·207	4·219	4·231	1 2 4	5 6 7	8 10
18	4·243	4·254	4·266	4·278	4·290	4·301	4·313	4·324	4·336	4·347	1 2 3	5 6 7	8 9
19	4·359	4·370	4·382	4·393	4·405	4·416	4·427	4·438	4·450	4·461	1 2 3	5 6 7	8 9
20	4·472	4·483	4·494	4·506	4·517	4·528	4·539	4·550	4·561	4·572	1 2 3	4 6 7	8 9
21	4·583	4·593	4·604	4·615	4·626	4·637	4·648	4·658	4·669	4·680	1 2 3	4 5 6	8 9
22	4·690	4·701	4·712	4·722	4·733	4·743	4·754	4·764	4·775	4·785	1 2 3	4 5 6	7 8
23	4·796	4·806	4·817	4·827	4·837	4·848	4·858	4·868	4·879	4·889	1 2 3	4 5 6	7 8
24	4·899	4·909	4·919	4·930	4·940	4·950	4·960	4·970	4·980	4·990	1 2 3	4 5 6	7 8
25	5·000	5·010	5·020	5·030	5·040	5·050	5·060	5·070	5·079	5·089	1 2 3	4 5 6	7 8
26	5·099	5·109	5·119	5·128	5·138	5·148	5·158	5·167	5·177	5·187	1 2 3	4 5 6	7 8
27	5·196	5·206	5·215	5·225	5·235	5·244	5·254	5·263	5·273	5·282	1 2 3	4 5 6	7 8
28	5·292	5·301	5·310	5·320	5·329	5·339	5·348	5·357	5·367	5·376	1 2 3	4 5 6	7 7
29	5·385	5·394	5·404	5·413	5·422	5·431	5·441	5·450	5·459	5·468	1 2 3	4 5 5	6 7
30	5·477	5·486	5·495	5·505	5·514	5·523	5·532	5·541	5·550	5·559	1 2 3	4 4 5	6 7
31	5·568	5·577	5·586	5·595	5·604	5·612	5·621	5·630	5·639	5·648	1 2 3	3 4 5	6 7
32	5·657	5·666	5·675	5·683	5·692	5·701	5·710	5·718	5·727	5·736	1 2 3	3 4 5	6 7
33	5·745	5·753	5·762	5·771	5·779	5·788	5·797	5·805	5·814	5·822	1 2 3	3 4 5	6 7
34	5·831	5·840	5·848	5·857	5·865	5·874	5·882	5·891	5·899	5·908	1 2 3	3 4 5	6 7
35	5·916	5·925	5·933	5·941	5·950	5·958	5·967	5·975	5·983	5·992	1 2 2	3 4 5	6 7
36	6·000	6·008	6·017	6·025	6·033	6·042	6·050	6·058	6·066	6·075	1 2 2	3 4 5	6 7
37	6·083	6·091	6·099	6·107	6·116	6·124	6·132	6·140	6·148	6·156	1 2 2	3 4 5	6 7
38	6·164	6·173	6·181	6·189	6·197	6·205	6·213	6·221	6·229	6·237	1 2 2	3 4 5	6 6
39	6·245	6·253	6·261	6·269	6·277	6·285	6·293	6·301	6·309	6·317	1 2 2	3 4 5	6 6
40	6·325	6·332	6·340	6·348	6·356	6·364	6·372	6·380	6·387	6·395	1 2 2	3 4 5	6 6
41	6·403	6·411	6·419	6·427	6·434	6·442	6·450	6·458	6·465	6·473	1 2 2	3 4 5	5 6
42	6·481	6·488	6·496	6·504	6·512	6·519	6·527	6·535	6·542	6·550	1 2 2	3 4 5	5 6
43	6·557	6·565	6·573	6·580	6·588	6·595	6·603	6·611	6·618	6·626	1 2 2	3 4 5	5 6
44	6·633	6·641	6·648	6·656	6·663	6·671	6·678	6·686	6·693	6·701	1 1 2	3 4 4	5 6
45	6·708	6·716	6·723	6·731	6·738	6·745	6·753	6·760	6·768	6·775	1 1 2	3 4 4	5 6
46	6·782	6·790	6·797	6·804	6·812	6·819	6·826	6·834	6·841	6·848	1 1 2	3 4 4	5 6
47	6·856	6·863	6·870	6·877	6·885	6·892	6·899	6·907	6·914	6·921	1 1 2	3 4 4	5 6
48	6·928	6·935	6·943	6·950	6·957	6·964	6·971	6·979	6·986	6·993	1 1 2	3 4 4	5 6
49	7·000	7·007	7·014	7·021	7·029	7·036	7·043	7·050	7·057	7·064	1 1 2	3 4 4	5 6
50	7·071	7·078	7·085	7·092	7·099	7·106	7·113	7·120	7·127	7·134	1 1 2	3 4 4	5 6
51	7·141	7·148	7·155	7·162	7·169	7·176	7·183	7·190	7·197	7·204	1 1 2	3 4 4	5 6
52	7·211	7·218	7·225	7·232	7·239	7·246	7·253	7·259	7·266	7·273	1 1 2	3 3 4	5 6
53	7·280	7·287	7·294	7·301	7·308	7·314	7·321	7·328	7·335	7·342	1 1 2	3 3 4	5 5
54	7·348	7·355	7·362	7·369	7·376	7·382	7·389	7·396	7·403	7·409	1 1 2	3 3 4	5 5

	0	1	2	3	4	5	6	7	8	9	Differences 1 2 3	4 5 6	7 8 9
55	7·416	7·423	7·430	7·436	7·443	7·450	7·457	7·463	7·470	7·477	1 1 2	3 3 4	5 5 6
56	7·483	7·490	7·497	7·503	7·510	7·517	7·523	7·530	7·537	7·543	1 1 2	3 3 4	5 5 6
57	7·550	7·556	7·563	7·570	7·576	7·583	7·589	7·596	7·603	7·609	1 1 2	3 3 4	5 5 6
58	7·616	7·622	7·629	7·635	7·642	7·649	7·655	7·662	7·668	7·675	1 1 2	3 3 4	5 5 6
59	7·681	7·688	7·694	7·701	7·707	7·714	7·720	7·727	7·733	7·740	1 1 2	3 3 4	5 5 6
60	7·746	7·752	7·759	7·765	7·772	7·778	7·785	7·791	7·797	7·804	1 1 2	3 3 4	4 5 6
61	7·810	7·817	7·823	7·829	7·836	7·842	7·849	7·855	7·861	7·868	1 1 2	3 3 4	4 5 6
62	7·874	7·880	7·887	7·893	7·899	7·906	7·912	7·918	7·925	7·931	1 1 2	3 3 4	4 5 6
63	7·937	7·944	7·950	7·956	7·962	7·969	7·975	7·981	7·987	7·994	1 1 2	3 3 4	4 5 6
64	8·000	8·006	8·012	8·019	8·025	8·031	8·037	8·044	8·050	8·056	1 1 2	2 3 4	4 5 6
65	8·062	8·068	8·075	8·081	8·087	8·093	8·099	8·106	8·112	8·118	1 1 2	2 3 4	4 5 6
66	8·124	8·130	8·136	8·142	8·149	8·155	8·161	8·167	8·173	8·179	1 1 2	2 3 4	4 5 5
67	8·185	8·191	8·198	8·204	8·210	8·216	8·222	8·228	8·234	8·240	1 1 2	2 3 4	4 5 5
68	8·246	8·252	8·258	8·264	8·270	8·276	8·283	8·289	8·295	8·301	1 1 2	2 3 4	4 5 5
69	8·307	8·313	8·319	8·325	8·331	8·337	8·343	8·349	8·355	8·361	1 1 2	2 3 4	4 5 5
70	8·367	8·373	8·379	8·385	8·390	8·396	8·402	8·408	8·414	8·420	1 1 2	2 3 4	4 5 5
71	8·426	8·432	8·438	8·444	8·450	8·456	8·462	8·468	8·473	8·479	1 1 2	2 3 4	4 5 5
72	8·485	8·491	8·497	8·503	8·509	8·515	8·521	8·526	8·532	8·538	1 1 2	2 3 3	4 5 5
73	8·544	8·550	8·556	8·562	8·567	8·573	8·579	8·585	8·591	8·597	1 1 2	2 3 3	4 5 5
74	8·602	8·608	8·614	8·620	8·626	8·631	8·637	8·643	8·649	8·654	1 1 2	2 3 3	4 5 5
75	8·660	8·666	8·672	8·678	8·683	8·689	8·695	8·701	8·706	8·712	1 1 2	2 3 3	4 5 5
76	8·718	8·724	8·729	8·735	8·741	8·746	8·752	8·758	8·764	8·769	1 1 2	2 3 3	4 5 5
77	8·775	8·781	8·786	8·792	8·798	8·803	8·809	8·815	8·820	8·826	1 1 2	2 3 3	4 4 5
78	8·832	8·837	8·843	8·849	8·854	8·860	8·866	8·871	8·877	8·883	1 1 2	2 3 3	4 4 5
79	8·888	8·894	8·899	8·905	8·911	8·916	8·922	8·927	8·933	8·939	1 1 2	2 3 3	4 4 5
80	8·944	8·950	8·955	8·961	8·967	8·972	8·978	8·983	8·989	8·994	1 1 2	2 3 3	4 4 5
81	9·000	9·006	9·011	9·017	9·022	9·028	9·033	9·039	9·044	9·050	1 1 2	2 3 3	4 4 5
82	9·055	9·061	9·066	9·072	9·077	9·083	9·088	9·094	9·099	9·105	1 1 2	2 3 3	4 4 5
83	9·110	9·116	9·121	9·127	9·132	9·138	9·143	9·149	9·154	9·160	1 1 2	2 3 3	4 4 5
84	9·165	9·171	9·176	9·182	9·187	9·192	9·198	9·203	9·209	9·214	1 1 2	2 3 3	4 4 5
85	9·220	9·225	9·230	9·236	9·241	9·247	9·252	9·257	9·263	9·268	1 1 2	2 3 3	4 4 5
86	9·274	9·279	9·284	9·290	9·295	9·301	9·306	9·311	9·317	9·322	1 1 2	2 3 3	4 4 5
87	9·327	9·333	9·338	9·343	9·349	9·354	9·359	9·365	9·370	9·375	1 1 2	2 3 3	4 4 5
88	9·381	9·386	9·391	9·397	9·402	9·407	9·413	9·418	9·423	9·429	1 1 2	2 3 3	4 4 5
89	9·434	9·439	9·445	9·450	9·455	9·460	9·466	9·471	9·476	9·482	1 1 2	2 3 3	4 4 5
90	9·487	9·492	9·497	9·503	9·508	9·513	9·518	9·524	9·529	9·534	1 1 2	2 3 3	4 4 5
91	9·539	9·545	9·550	9·555	9·560	9·566	9·571	9·576	9·581	9·586	1 1 2	2 3 3	4 4 5
92	9·592	9·597	9·602	9·607	9·612	9·618	9·623	9·628	9·633	9·638	1 1 2	2 3 3	4 4 5
93	9·644	9·649	9·654	9·659	9·664	9·670	9·675	9·680	9·685	9·690	1 1 2	2 3 3	4 4 5
94	9·695	9·701	9·706	9·711	9·716	9·721	9·726	9·731	9·737	9·742	1 1 2	2 3 3	4 4 5
95	9·747	9·752	9·757	9·762	9·767	9·772	9·778	9·783	9·788	9·793	1 1 2	2 3 3	4 4 5
96	9·798	9·803	9·808	9·813	9·818	9·823	9·829	9·834	9·839	9·844	1 1 2	2 3 3	4 4 5
97	9·849	9·854	9·859	9·864	9·869	9·874	9·879	9·884	9·889	9·894	1 1 2	2 3 3	4 4 5
98	9·899	9·905	9·910	9·915	9·920	9·925	9·930	9·935	9·940	9·945	0 1 1	2 2 3	3 4 4
99	9·950	9·955	9·960	9·965	9·970	9·975	9·980	9·985	9·990	9·995	0 1 1	2 2 3	3 4 4

Printed in the United States
By Bookmasters